The Delectable Vegetable

The Delectable Vegetable

by KAY SHAW NELSON

The John Day Company

Copyright © 1976 by Kay Shaw Nelson

All rights reserved. Except for use in a review, the reproduction or utilization of this work in any form or by any electronic, mechanical, or other means, now known or hereafter invented, including xerography, photocopying, and recording, and in any information storage and retrieval system is forbidden without the written permission of the publisher. Published simultaneously in Canada by Fitzhenry & Whiteside Limited, Toronto.

Designed by Lynn Braswell

Manufactured in the United States of America

Library of Congress Cataloging in Publication Data

Nelson, Kay Shaw.
  The delectable vegetable.

  1. Cookery (Vegetables) I. Title.
TX801.N44   641.5'5   76-3728
ISBN 0-381-98292-0

1 2 3 4 5 6 7 8 9 10

To My Sister,
Mary Shaw Berry,
a Dedicated Vegetable Cook

Contents

Introduction   ix
About Vegetable Cookery   xiii
Appetizers   1
Savory Soups   19
Stuffed Vegetables   37
Main Dishes   56
Especially for Company   76
Greens & Leafy Vegetables   93
Dried Vegetables   110
Vegetables as Accompaniments   126
Salads   141
Curious Vegetables   155
Breads & Desserts   171
Pickles, Relishes, & Sauces   185
Index   200

Introduction

When I was a child in New Hampshire, I always looked forward to summer. Partly, of course, this was due to the prospect of release from the rigors of the hard cold winter. Equally pleasurable to me, however, was the promise of summer gardening and the bounty of delectable vegetables this would bring to our family table.

Year after year it seemed that the green beans, squash, peas, lettuce, radishes, and other vegetables we grew tasted better than anything else. Even a simple golden carrot pulled from the earth, with the dirt freshly wiped off, was a tasty treat. At the end of a long day of hoeing and weeding, herb-sprinkled cucumbers were welcome appetizers. And the special day when the first corn ripened was a Lucullan event.

That the variety of these vegetables was limited did not detract in the slightest from my enjoyment of them. We never grew too many kinds, and there were no exotic or unusual choices. Even so, I can still recall those freshly sliced tomatoes on mayonnaise-covered homemade bread on which we lunched for days in a row, and have never found a plain new potato, simply garnished with butter, which could compare with those grown by my father.

Our family preoccupation with vegetables was a year-round pastime. During the winter we would look through seed catalogs and periodicals from mail-order houses to see if there were any new kinds or different varieties developed since the previous season. Usually this was merely a way of amusing ourselves, for we rarely purchased from them or increased our range of vegetables. We could get garden staples from the local store, and each spring we planted the same familiar New England standbys that we knew would thrive in our northern climate.

The spring garden duties never seemed like chores but were welcomed as forms of exercise, a way of getting fresh air, having some fun, and doing something that needed to be done—all at one and the same time. Planting a garden is indeed hard work, but it is also rewarding. As you ardently clean and prepare the ground, there are visions of the marvelous meals that will eventually result from your dedicated efforts. Springtime was also when we gathered young greens from the fields, and our winter-weary palates welcomed their clean fresh taste.

Throughout the summer there was, of course, the continuous duty of weeding, but more pleasant, was the constant lookout for the first Swiss chard, beet greens, asparagus stalks, and peas, so that we could have them as young as possible for our meals. Generally the cooking of our various garden vegetables, as they became available, was quite simple, so that the natural flavor could be fully appreciated and the nutrients preserved. Seasonings were primarily herbs, perhaps some onion, salt and pepper. A few kinds were served in a cream sauce, and we also had nourishing soups, chowders, salads, baked dishes, and on occasion, deep-fried treats, such as fritters and croquettes. Such was the goodness of these dishes that during the summer season we scarcely touched meat and fish because the bounty from our garden was too tempting to forego.

In late summer our kitchen was always a beehive of activity, for we had to put up a good supply of vegetables for the winter. In those days, prior to freezing, our foods were primarily canned; a few, like shelled beans, could be dried. But in August, both our wood-burning black iron stove and the oil stove groaned with the weight of large kettles filled with carefully packed pints and quarts simmering away. There was also pickling and preserving, and later, as the garden was cleaned for fall, we dug the potatoes and other root vegetables to store in the cellar.

By Thanksgiving the vegetable garden was bare and covered with frost, or more often snow, and our family's holiday meal was a proud parade of dishes starring the product of our labors. We dined on fluffy mashed potatoes, creamed onions, succotash, maple-glazed butternut squash, butter-flavored turnips, cucumber pickles, steamed carrot pudding, and rich pumpkin pie, along with other traditional fare.

And then throughout the winter we trekked daily to the basement cold-storage room and root cellar for staples with which to prepare such body-warming and rib-sticking dishes as corn chowder, beet hash, New England boiled dinner, Boston baked beans, parsnip stew, stuffed onions, pot roast, and chicken pie with vegetables and dumplings.

Since those days of my childhood, I have discovered in other American locales, as well as in foreign countries, wider horizons to the pleasures of vegetable cookery and dining. In fact, I have been faintly surprised to find that the enjoyment of a wide assortment of these particular foods is timeless and cosmopolitan, and that much can be learned about their preparation from other cuisines, including those of our own land.

I have probed the cookery of the past and learned what our forefathers did with Jerusalem artichokes, oyster plant, mustard greens, lamb's-quarters, and pumpkin. In Pennsylvania I discovered the glories of sauerkraut, unusual potato dishes, colorful pickles and relishes, and sweet-sour flavorings. In the Southwest I was introduced to a fascinating array of dishes made with corn, beans, and peppers, both sweet and hot, which owe much to the cookery of their Latin American neighbors.

Southerners, I found, were experts in combining their favorite okra, sweet potatoes, black-eyed peas, collard greens, and tomatoes with deft seasonings and other foods to make inspiring fare. In the Midwest I discovered innovative corn preparations and was delighted with a repertoire of dishes brought from Scandinavia. I soon realized that the Far West was fortunate in having varied artichoke dishes and marvelous salads.

I found the differences between the various cuisines intriguing. Yet this was only the beginning of my exploration of the vegetable kingdom. In the Middle East, where many of our familiar plants had their origins, I whiled away my spare time visiting open-air street markets to view handsome displays of garden-fresh foods. There were new acquaintances like fava beans; old friends in new forms, such as slender foot-long cucumbers and round eggplants; and others I had cooked less often like leeks and red peppers.

In Europe I never tired of sampling each country's characteristic vegetable dishes. Germans do wonders with the gnarled and knobby brown celery root and the almost forgotten kohlrabi. I can still recall the delight of sampling fennel salad and savory baked zucchini in Italy, and the tender artichokes of Rome, prepared in so many ways that even ice creams are made with them.

In Spain the Basque dishes were carefully made with tomatoes, green peppers, and herbs. In Belgium endive and mushrooms prepared in various ways added allure to my meals—and once a year the marvelous white asparagus was a treat without equal.

While I lived in the Orient, friends from many lands introduced me to age-old recipes for preparing spicy relishes, such as the hot Korean *kimchee*, the long and firm white radish, bean sprouts and soy beans, celery cabbage, and various roots.

In recent years it has been a fruitful experience to learn how the cooks of the various cuisines of Africa, the Caribbean, and Latin America rely on vegetables which, at first seem quaint but are actually staples, like the yam and *akee;* the chayote, a member of the squash family; the *calabaza*, West Indian pumpkin; the small pear-shaped *christophene;* and extra-long string beans.

Dining and cooking around the world provided me with a wide number of intriguing members of the vegetable kingdom. From the kitchens and restaurants of the various countries I have collected a representative selection of recipes, which display the goodness and versatility of these foods.

Now it is a pleasure to witness a welcome renaissance in vegetable cookery in our country. Too often in the recent past, these valuable plants were given short shrift by the cook, poorly treated and prepared in the same old way over and over again. The vegetable fare in America has been noted for its dullness and limitations.

Fortunately there is no magic or mystique about vegetable cookery. All one needs is a desire to broaden one's knowledge of the individual kinds and their preparation, some time and effort, and most important, a bit of imagination.

To assist in the cooking of these cosmopolitan vegetables, and the dishes made with them, some basic data is offered in the forthcoming pages. As for the recipes, some of the ingredients and the manner of some preparation will be familiar; others will not. The book itself was devised and written in dedication to the cooks of the world who had the artistry to create the recipes. We are fortunate that this rich heritage is available to us.

About Vegetable Cookery

As there are more than three hundred varieties of edible vegetables in the world, man has a wide range of plants, roots, and herbs from which to choose. We have leafy, stem, tuber, flower, seed, pod, and even fruits that are classified as vegetables. Some are eaten raw, some cooked, others in either way.

Whereas Americans once had to rely on the fresh vegetables grown in their own particular area and were limited to seasonal choices, we now have a plethora of varieties from around our own country, and abroad that are generally available throughout the year. Furthermore, there are excellent frozen and canned vegetables.

Unfortunately, most Americans have not taken advantage of the great opportunity that these treasures offer. It is a rare cook who would recognize, let alone know what to do with, celeriac, oyster plant, Jerusalem artichokes, even parsnips, and yet these were once staples on the family table. Too often the selection is limited to a few well-known vegetables that are cooked and served in a mundane fashion and are given no distinctive place on the menu.

To enjoy vegetable cookery, it is advisable to increase the range of your knowledge of those that are not only easily available but can be assets for family meals and company entertaining. Gradually, others can be added to the repertoire. Go to a vegetable market or supermarket and pick out any unfamiliar or strange-looking choice. Find out its name, take it home, and learn what to do with the new acquisition. It's fun to prepare and serve a new food.

Properly prepared vegetables are very important to the pleasure of daily meals and can add allure to company repasts. They are good meat and poultry substitutes for some occasions. Although a few are regarded merely as garnishes or supplements to other foods, each vegetable can occupy an important place on the menu, and therefore should be held in high esteem and accorded special treatment. Vegetables can provide nourishing, inexpensive, and attractive dishes, ranging from appetizers to desserts.

They are essential to the diet because of their importance as a source of vitamins B and C and of carotene—a precursor of vitamin A. They are also rich in minerals, particularly iron and calcium, and have varying amounts of thiamine, riboflavin, and niacin. Vegetables also have a low calorie value.

It is not enough, however, to think of vegetables only as good for us. They are also attractive and delectable, and should not be limited to a subordinate or ignominious role on our menus. Marvelous vegetable soufflés, pies, puddings, casseroles, stews, and stuffed vegetables can be starred as inviting main courses.

Cooks around the world have long paid homage to our many vegetables by preparing them in delightfully different ways. Fortunately, their knowledge and culinary expertise have been passed on to modern cooks, who should be able to find great pleasure in enhancing the art of vegetable cookery.

Buying Vegetables

Do not hastily purchase fresh vegetables but take time to look them over and choose the best. Indications of goodness and freshness are color, texture, size, and age. They should be well-colored, crisp or firm, free of blemishes and bruises, and ripe but not overmature. The texture of an overripe vegetable is often stringy or woody and cannot be corrected by cooking. Very often the smaller the vegetable, the tastier it will be. Today, in a few specialized markets, there are inviting midget vegetables.

During the summer months vegetables are at their peak in

quality and quantity and are less expensive. It is unwise, however, to buy more than you can use as vegetables start to deteriorate as soon as they are picked, and unless properly cared for and stored, will lose much of their appeal and nutritive value. Therefore, use them as soon as possible after purchase.

Most vegetables will keep for short periods in the refrigerator. Potatoes, carrots, beets, onions, turnips, and other root vegetables can be stored in a cool, dry dark room where there is a circulation of air.

Preparing and Cooking Vegetables

Nature imparts natural goodness to all parts of the vegetable, and the cook should strive to retain it. In order to preserve the shape, texture, color, and flavor of each vegetable, consider the following suggestions:

1. Prepare vegetables shortly before cooking as air exposure detracts from food values and appearance.
2. Do not peel vegetables before cooking unless necessary. The greatest concentration of vitamins and minerals is near the skin so it is best to remove it after cooking.
3. Do not soak vegetables in water before cooking as the nutritive elements are soluble in water and are lost when it is discarded.
4. For the same reason, use no water or as little as possible when cooking.
5. Certain vegetables such as artichokes and oyster plant darken when they are prepared and exposed to the air. To prevent this, drop into water to which a little vinegar or lemon juice has been added.
6. Never use baking soda to preserve the color of vegetables while they are cooking as it destroys some of the vitamins and also spoils the texture and flavor of the vegetables. Lemon juice can be added to vegetables that might discolor while they are cooking.
7. Never overcook vegetables. It is better to have them undercooked. Cooked until just tender is the preferable way.
8. If vegetables cannot be served at once, it is better to reheat them than keep them warm.

Boiling Vegetables This most common method of cooking vegetables can be done in a small amount of water or other liquid. Use as little as possible, just enough to prevent scorching. Such a vegetable as spinach can be cooked with only the water left clinging to its leaves after they have been washed. It is also good to have uniform pieces of equal size so the vegetable will cook evenly and all of it be done at the same time. The liquid should be boiling when the vegetables are added, and then cooked quickly over medium heat. Avoid violent boiling and do not stir. Most vegetables should be boiled tightly covered; the exceptions are cabbage, cauliflower, beets, onions, turnips, and parsnips, which are best cooked uncovered.

To blanch vegetables, such as whole green peppers to be stuffed, parboil in boiling water 3 to 4 minutes. Drain.

Braising Vegetables This method of cooking slowly in moist heat in a tightly covered pan over low heat is excellent for such vegetables as carrots, onions, turnips, etc. For liquid, use a small quantity of water or consommé.

Frying Vegetables A great number of vegetables are excellent fried. Some may be cooked whole; others cut up. Some are fried raw; others boiled beforehand. In many cases the vegetables are first coated or dipped into a batter. There are two methods of frying: one is in a small amount of fat in a skillet; the other is in a larger amount of fat in a kettle or deep-fat fryer. Before frying, be sure that your vegetables are dry as water makes the fat splatter. For deep-fat frying, the temperature can vary, according to the vegetable, from 345 to 375°F. Cook quickly so the vegetable will be crisp on the outside and moist within. Drain on absorbent paper, season with salt, and serve at once.

Steaming Vegetables Steamed vegetables retain their appearance, shape, and original flavor, as well as their nutrients. A longer cooking time, however, is required. To steam vegetables, fill a large saucepan or kettle with water, bring the water to a boil, place the vegetables in a strainer or vegetable container over it, and cook until just tender.

Appetizers

Our varied and colorful vegetables make appealing appetizers. Whether served raw or cooked, alone or in combination with other foods, they are tempting to the eye, pleasing to the palate, and stimulating to the appetite.

Today's cook, ever mindful of time and budget when preparing culinary creations, can profit by serving vegetables with drinks before a meal or at a party, as well as for a first course at the table. For most of the appetizers featuring vegetables can be prepared beforehand, are inexpensive, and while providing necessary preprandial sustenance, offer excellent nutritional value and yet are low in calories.

Cold and hot vegetable appetizers have long been favorite fare at before-the-meal gatherings around the world. Many of the age-old preparations and ways of serving them are as good now as they were in yesteryear.

Europeans, in particular, created imaginative galaxies of appetizers in which vegetables were prominently and attractively displayed. A Scandinavian *smorgasbord* still includes anchovy-flavored potato dishes and mixed vegetable salads. For the richly embellished Russian *zakuski*, there are eggplant and mushroom "caviars," radishes in sour cream, bean dishes, and a plethora of pickles. The French *hors d'oeuvre* and Italian *antipasti* star a wide variety of well-seasoned raw, cooked, and pickled vegetables.

Of all the world's diverse vegetable appetizers, particularly appealing are those featuring an assortment of raw vegetables. In France, for example, it is the custom to serve cold appetizers before luncheon. The *crudités,* or raw vegetables, are among the best of the delicious array. The selection might include paper-thin cucumber slices, finely grated raw carrots, sliced celery stalks, baby artichokes, strips of fennel, wedges of green pepper, tomato slices, whole red and white radishes, and cauliflower florets, each kind individually bathed in herb-flavored sauces or spicy marinade and served in small glass dishes.

Crudités can easily be prepared at home by adorning a few or several raw vegetables with homemade or purchased salad dressings ranging from French or Italian to an oil-vinegar-herb marinade.

A favorite way of serving both raw and cooked vegetables in Italy's northern Piedmont region is to dip them in a rich truffle-flavored cream sauce called *bagna cauda,* or hot bath. In French Provence a similar specialty is made with a rich and potent garlic sauce, *aïoli.*

There is no end to the possibilities for preparing vegetables as appetizers. For entertaining, they are welcomed by guests weary of the sameness of our cocktail and buffet fare that has long relied on cold cooked meats and seafood. Even the simple red radish with topping of green intact, presented with a pat of butter as it is in France, will become a conversation piece.

Considerable thought should be given to the selection of appetizers as they set the stage, so to speak, for the culinary event. Provide a light and attractive appetizer if the meal is to include a number of courses and supply a more substantial selection before a lighter repast. Serve the vegetables in attractive dishes and, if necessary, garnish them.

Included in this chapter is a collection of inviting appetizers, culled from around the world, that can be used for family meals or for entertaining.

Oriental Bean-Sprout Appetizer

Delicious crisp bean sprouts—young shoots of a variety of soybean, a great source of protein—are widely available in cans or occasionally fresh in some markets. They also can easily be grown in the home, by sprouting tiny mung beans, or peas.

1 large onion, peeled	2 cups (1 pound can) bean sprouts, drained
1 garlic clove, crushed	
2 tablespoons peanut or vegetable oil	$1/3$ cup sliced water chestnuts
	3 to 4 tablespoons soy sauce
1 large green pepper, cleaned and cut into slivers	Pepper to taste

Cut onion in half from top to bottom; slice thinly. Sauté with garlic in heated oil in a medium saucepan until just tender. Add green pepper and sauté 1 minute. Mix in bean sprouts and water chestnuts, and sauté until hot. Add soy sauce and pepper. Heat 1 minute. Serves 4.

Persian Spinach Borani

This traditional Persian, or Iranian, appetizer can also be served as a salad or as an accompaniment to meat or poultry. Raw cucumbers, cooked mushrooms, or eggplant may be substituted for the spinach, if desired.

2 packages (10 ounces each) frozen spinach	2 tablespoons butter or vegetable oil
1 large onion, peeled and finely chopped	1 cup plain yogurt, about
	$1/2$ teaspoon ground cinnamon
1 or 2 garlic cloves, crushed	Salt and pepper to taste

Cook spinach according to package directions; drain well. Press out all liquid with back of a spoon; chop finely or whirl in a blender. Set aside. Sauté onion and garlic in butter in a saucepan until tender. Mix in spinach and sauté about 2 minutes to blend ... and pepper; remove from heat.

Chili Bean Dip from Mexico

This inexpensive dip is made with pale pink and brown-speckled dried beans, called *pinto* by early Spanish explorers after their word for paint. The beans turn red-brown during cooking and are very nutritious.

- 1 can (about 1 pound) pinto or red beans, drained
- 1 medium-sized onion, peeled and minced
- 2 tablespoons vegetable oil
- ¼ teaspoon dried oregano
- 1 to 2 tablespoons chili powder
- 1 can (8 ounces) tomato sauce
- Salt and pepper to taste

Mash beans. Sauté onion in heated oil until tender. Add oregano and chili powder, and cook 1 minute. Mix in tomato sauce, salt and pepper, and cook slowly, uncovered, for 10 minutes. Add mashed beans and cool. Serve with corn chips or Fritos. Serves about 8.

Provençal Marinated Carrots

The colorful carrot, an excellent source of vitamin A, has been developed from an ancient weed to a vegetable of many varieties, most of them quite large. The best and most flavorful are the "baby" carrots, just two to three inches long, and preferably freshly picked. They are also available in cans.

- 1½ cups cooked baby carrots, or 1 can (about 1 pound) baby carrots, drained
- 6 tablespoons olive or vegetable oil
- 2 tablespoons tarragon or wine vinegar
- 1 teaspoon sugar
- ¼ teaspoon dried basil
- Salt and pepper to taste
- 4 scallions, cleaned and chopped
- 1 or 2 garlic cloves, crushed
- 3 medium-sized tomatoes, peeled and chopped
- 3 tablespoons chopped fresh parsley

Put carrots in a bowl and add 5 tablespoons of oil, the vinegar, sugar, basil, salt and pepper. Refrigerate, covered, overnight or for about 8 hours, turning carrots occasionally. Sauté scallions and garlic in 1 tablespoon heated oil in a small skillet until tender. Add tomatoes and sauté 5 minutes. Mix in parsley. Season with salt and pepper. Remove carrots with a slotted spoon from marinade and put on a plate. Top with the tomato mixture. Serves 4.

Asparagus Canapés from Germany

An easy-to-prepare and elegant appetizer.

Butter thin slices of rye bread or pumpernickel cut into 1- by 3-inch rectangles. Place on each a cooked fresh or frozen green asparagus spear. Sprinkle with a little fresh lemon juice, dried thyme, salt and pepper. Pipe or spoon mayonnaise along the top and around the edges. Serve at once or refrigerate, covered with wax paper, until ready to serve.

Note: Canned asparagus can also be used.

Indian Vegetable Pachadi

This appetizer from southern India may be served in individual small bowls and eaten with spoons or served in a large bowl as a dip, accompanied by *chapati*—unleavened Indian bread—or crackers.

1 medium-sized onion, peeled and minced	1 teaspoon minced hot chilis (optional)
1 garlic clove, crushed	1 cup diced peeled cucumber
2 tablespoons vegetable oil	2 cups plain yogurt
1 to 2 tablespoons curry powder	Salt and pepper to taste
1 to 2 teaspoons chili powder	2 tablespoons chopped fresh coriander or parsley
2 medium-sized tomatoes, peeled and chopped	

Sauté onion and garlic in heated oil in a skillet until tender. Add curry and chili powders, and cook 1 minute. Remove from heat and mix with remaining ingredients. Cool. Serves 6 to 8.

French Green Pea Croquettes

These small appealing croquettes from France are made with green peas, preferably the tiny ones known as *petits pois*, picked before they reach maturity and noted for their sweet and tender quality. Garden or green peas should be cooked very quickly, just until tender but firm. Mint is an excellent flavoring for peas.

- 2 tablespoons butter or margarine
- 2 tablespoons minced onions
- 2 cups puréed cooked green peas
- 3 tablespoons chopped fresh mint
- Salt and pepper to taste
- 3 eggs
- Fine dry bread crumbs
- 2 tablespoons milk
- Fat for frying

Melt butter in a small skillet and sauté onions in it until tender. Add peas, mint, salt and pepper, and mix well. Remove from stove and mix in 2 eggs, previously beaten a little. Spread evenly in a flat dish and chill. Shape mixture into 2-inch–long and about 3/4-inch–thick rectangles. Roll in bread crumbs. Dip in remaining beaten egg mixed with milk. Roll again in bread crumbs. Arrange on a wire rack or plate and chill. Leave at room temperature 1 hour before cooking. Fry croquettes in hot deep fat (370° F. on frying thermometer) until crisp and golden brown. Drain on absorbent paper. Serve as an appetizer or first course. Makes 16.

Belgian Fresh Mushroom-Herb Appetizer

Wild mushrooms, once called "food of the Gods," have been treasured since ancient times, but cultivated mushrooms, now sold widely in America, were first grown by the French during the seventeenth century. Mushrooms impart a rich and appealing flavor to a diverse selection of elegant dishes. Do not peel mushrooms or soak them in water. Handle them with care and cook them only a few minutes. Raw mushrooms make good appetizers and may be added to salads. The caps may be stuffed with cheese, seafood, or meat.

- 1 pound fresh mushrooms
- 3 to 4 tablespoons butter or margarine
- 2 tablespoons fresh lemon juice
- Salt and pepper to taste
- 1/3 cup chopped fresh herbs (parsley, tarragon, rosemary, dill)
- 3 tablespoons sour cream (optional)

Clean mushrooms by wiping quickly with wet paper toweling to remove any dirt. Cut off any woody stem ends. Slice thickly. Sauté in heated butter and lemon juice for 4 minutes. Add salt, pepper, and herbs. Spoon into a shallow dish and top with sour cream, if desired. Serve with small pieces of dark or white bread. Serves 4.

Zucchini Quiche au Fromage

Zucchini, an attractive green-and-yellow striped variety of summer squash developed in Italy, which is also known as Italian squash or vegetable marrow, should not be peeled before cooking. Just wash it and cut off the ends. Zucchini is an exceedingly versatile vegetable. Even its orange buds may be sautéed in oil and butter, and added to omelet or scrambled eggs.

- 2 medium-sized zucchini, about 1/2 pound each, washed and stemmed
- 1 medium-sized onion, peeled
- About 3 tablespoons olive or vegetable oil
- 1/2 teaspoon dried basil
- Salt and pepper to taste
- 4 eggs
- 1/2 cup sour cream at room temperature
- 1/2 cup grated Parmesan cheese
- Single 9-inch pastry shell, baked 10 minutes and cooled

Wipe zucchini dry and slice thinly. Cut onion in half from top to bottom and slice thinly. Sauté onion in heated oil in a skillet until tender. Push aside and add zucchini slices, several at a time, and sauté until tender, adding more oil if needed. Remove to a bowl when cooked. Add basil, salt and pepper, and mix well. In another bowl beat together eggs, sour cream, and cheese. Add to zucchini. Turn into pastry shell, spreading evenly. Bake in preheated 375° F. oven in middle level for about 30 minutes, until puffed and golden and a knife inserted into center comes out clean. Serves 6.

Near Eastern Cracked Wheat-Vegetable Appetizer

This dip, called *tabbouleh* in Syria and Lebanon, is made with nutty-flavored raw *bulgur*—cracked wheat—which is sold in specialty or health food stores. The preparation requires some chopping, but the appetizer is a sure-fire winner whenever served, either to a small group or a large party.

1 cup fine cracked wheat (bulgur)	2 teaspoons salt
2 cups chopped scallions, with tops	1 teaspoon pepper
	½ cup fresh lemon juice
2 cups chopped fresh parsley	About ⅓ cup olive or vegetable oil
¼ cup chopped fresh mint	
2 cups peeled chopped tomatoes	

Soak cracked wheat in water to cover for 30 minutes. Squeeze dry by pressing between palms of hands. Combine with remaining ingredients in a large bowl. Mix well and chill. Serve it in a bowl or shaped into a mound on a flat dish surrounded with fresh lettuce, cabbage, or grape leaves to be used as scoops, or with unleavened Arabic bread or crackers. Serves about 10.

German Sauerkraut Balls

Pickled shredded green cabbage, or sauerkraut, is an ancient food that originated in China and was brought to Europe by invading Tartars centuries ago. Now sold fresh in bulk, in cans, or in plastic bags, it once was commonly prepared in the home each autumn. Sauerkraut is widely eaten in Europe and the Orient as a healthful food, rich in vitamins, phosphorus, calcium, and iron. It deserves to be used more widely and creatively in dishes such as this one.

- 1 medium-sized onion, peeled and minced
- 2 tablespoons butter or margarine
- 2 cups finely ground or minced cooked ham
- 1 tablespoon prepared mustard
- Pepper to taste
- 2 cups finely chopped drained sauerkraut
- About $2/3$ cup all-purpose flour
- $1/2$ cup beef consommé
- 2 eggs, well beaten
- About $1/2$ cup fine dry bread crumbs
- Fat for deep frying

Sauté onion in heated butter until tender. Add ham and cook, stirring frequently, for 2 or 3 minutes. Mix in mustard, pepper, sauerkraut, $1/2$ cup flour, and consommé. Cook for 10 minutes. Spread out on a plate to cool. Shape into 1-inch balls and refrigerate 1 hour. Roll balls in flour, dip in beaten eggs, and roll in bread crumbs. Fry in hot deep fat (370° F. on frying thermometer) until golden brown. Drain on absorbent paper. Serve with toothpicks. Makes about 5 dozen.

Radishes in Sour Cream

This is an interesting way of serving the common red radish, full of minerals and vitamins A and C, and noted for its crisp texture. Radishes come in a number of shapes and may be white, black, purple, pink, or yellow, as well as red, in color. Usually eaten raw, they can also be cooked, when generally they are steamed. If tender, the radish tops can also be eaten, either raw or cooked.

2 hard-cooked eggs, shelled
1 cup sour cream
Salt and pepper to taste
2 cups sliced red radishes
2 tablespoons chopped fresh dill

Cut eggs crosswise into halves and remove yolks. Mash yolks; chop whites. Combine yolks with sour cream, salt and pepper. Put radishes in a small bowl and cover with sour-cream mixture. Sprinkle with chopped egg whites and dill. Chill. Serves 4.

Marinated Leeks, Romanian Style

Leeks, which are members of the onion family, have sweet-flavored thick white bulbs and large green leaves. The bulbs are edible, but the leaves must be cut off before cooking. Leeks have to be very carefully washed to remove all dirt. They are excellent cooked by themselves or in combination with other foods, especially vegetables.

6 medium-sized leeks
Boiling salted water
$1/3$ to $1/2$ cup olive or vegetable oil
Juice of 1 lemon
$1/4$ teaspoon dried thyme
$1/2$ teaspoon fennel seeds
1 bay leaf
Salt and pepper to taste
3 tablespoons chopped fresh dill or parsley
Garnish: tomato wedges and green pepper strips

Cut green tops from leeks. Wash white parts very well under running water to remove any sand. Cut into 1-inch lengths. Cook in a little boiling salted water until just tender, about 15 minutes. Drain well and put in a shallow dish. Cover with remaining ingredients and refrigerate overnight or for about 8 hours. Discard bay leaf. Serve garnished with raw tomato wedges and green pepper strips, if desired. Serves 4 to 6.

California Artichoke-Tomato Cocktail

The globe artichoke, a member of the daisy family and an aristocratic relative of the humble thistle, is a vegetable that deserves to be better known in America. It has an appealing nutlike flavor, and the leaves of whole cooked artichokes are eaten with the fingers, peeled off one at a time and dipped into melted butter or sauce. The innermost portion of the artichoke is called the heart. Hearts of artichoke, generally available in cans or frozen, are eaten with a fork.

1 package (9 ounces) frozen artichoke hearts	Salt and pepper to taste
2 medium-sized tomatoes, peeled and chopped	4 medium-sized lettuce leaves, torn into shreds
1 cup chopped celery	About ¼ cup mayonnaise or salad dressing
2 tablespoons fresh lemon juice	2 tablespoons chopped fresh dill
¼ teaspoon dried thyme	

Cook artichoke hearts according to package directions. Drain and cool. Combine with tomatoes, celery, lemon juice, thyme, salt and pepper. Chill. When ready to serve, arrange lettuce leaves in 4 small serving dishes. Spoon artichoke-tomato mixture over them and garnish with a spoonful of mayonnaise, or salad dressing, and dill. Serves 4.

Peruvian Sweet-Potato Appetizers

Despite their name, the white and the sweet potato have nothing in common. The former is a member of the nightshade family, the latter of the morning-glory family. The sweet potato has long been treasured as a staple food in tropical areas of the Americas, where it originated. It should be cooked, generally baked or boiled, without peeling to preserve nutrients.

2½ cups mashed, cooked, or canned sweet potatoes	1 egg, beaten
1 tablespoon all-purpose flour	¼ teaspoon ground nutmeg
1 to 2 tablespoons brown sugar	¼ cup vegetable oil
1 tablespoon melted butter or margarine	

Combine all ingredients, except oil, in a bowl. Chill in refrigerator. Just before serving, heat oil in a skillet. Drop potatoes by spoonfuls into hot fat and cook on both sides, turning once or twice. Serve hot. Serves about 10.

Russian Mushroom "Caviar"

In Russia this popular appetizer has been dubbed "poor man's caviar," as mushrooms are more widely available than expensive "black pearls," or sturgeon caviar.

1 cup scallions, with some tops	Salt and pepper to taste
¼ cup butter or margarine	3 tablespoons chopped fresh dill or parsley
2 cups chopped fresh mushrooms, previously cleaned	⅔ cup sour cream
2 tablespoons fresh lemon juice	

Sauté scallions in butter in a small skillet for 1 minute. Add mushrooms, lemon juice, salt and pepper. Sauté 4 minutes. Remove from heat and stir in dill and sour cream. Serve warm or cold with small pieces of black bread or pumpernickel. Serves 4.

The Delectable Vegetable

Miniature Potato Balls from Turkey

A simple but appealing appetizer that can be prepared beforehand and cooked just prior to serving.

4 medium-sized potatoes	Salt and pepper, to taste
Boiling salted water	About ½ cup all-purpose flour
2 tablespoons butter or margarine	About ½ cup vegetable oil or shortening
1 egg yolk	
½ cup grated Parmesan cheese	
2 tablespoons chopped fresh parsley	

Scrub potatoes and cook in boiling salted water until tender, about 20 minutes. Drain. Return to stove and leave over low heat for several seconds, shaking almost constantly to dry well. Peel, and while still warm, mash and beat until light and fluffy. Add butter, egg yolk, cheese, and parsley. Season with salt and pepper. Mix well. Shape into 1-inch balls. Roll in flour and refrigerate 2 hours. Heat about ⅛ inch oil or shortening in a skillet and fry potato balls in the hot fat, turning to cook on all sides, until golden brown. Drain on absorbent paper. Serve hot. Makes about 35.

English Cottage Cheese-Vegetable Dip

A good low-calorie and nutritious appetizer.

2 cups creamed cottage cheese	⅓ cup sliced radishes
¼ cup sour cream or plain yogurt	Salt and pepper to taste
⅓ cup minced raw carrots	3 tablespoons chopped fresh dill or parsley
⅓ cup chopped green peppers	

Combine all ingredients, except dill. Chill in refrigerator. Just before serving, garnish with dill. Makes about 3 cups.

Cucumber Boats from France

The fruit of a vine belonging to the gourd family, the cucumber can be very thick and short or slender and long, sometimes as long as two feet. Raw cucumbers are excellent for making canapés, sandwiches, or shells for filling. Cucumbers are also good when cooked or pickled.

>   2 medium-sized cucumbers
>   2 cups flaked cooked salmon or tuna fish
>   1/4 cup minced celery
>   About 1/4 cup mayonnaise
>   4 teaspoons sharp mustard
>   Juice of 1/2 lemon
>   1/4 teaspoon dried rosemary
>   Salt and pepper to taste

Peel cucumbers and cut in halves lengthwise. Scoop out seeds and pulp with a small spoon or grapefruit knife, and discard them. Combine remaining ingredients and spoon into cucumber shells. Refrigerate until ready to serve. Cut into 1- or 1½-inch lengths to facilitate serving, if desired. Serves 8 to 10.

Riviera Eggplant Appetizer

The handsome shiny eggplant, grown in several sizes and shapes, often has a dark purple skin, but it can also be white, yellowish, red, or even striped. Actually a fruit but widely regarded as a vegetable, the eggplant was once thought to be poisonous and was shunned as a food. It has long been treasured, however, in Mediterranean countries, where it is often cooked with garlic, onions, olive oil, cheeses, fresh and dried herbs, and tomatoes, as well as other vegetables.

>   1 medium-sized eggplant, about 1¼ pounds, washed
>   6 scallions, with tops, minced
>   2 garlic cloves, crushed
>   2 medium-sized green peppers, cleaned and chopped
>   2 large tomatoes, peeled and chopped
>   3 tablespoons fresh lemon juice
>   1/3 to 1/2 cup olive oil
>   Salt and pepper to taste

Prick eggplant in several places and put on a cookie sheet. Bake in a preheated 400° F. oven until soft, about 50 minutes. Cool a little. Peel off and discard skin. Put pulp in a bowl; chop with a wooden spoon or knife. Pour off all liquid. Add remaining ingredients and mix well. Cool. Serve in a bowl, or as a mound on a plate, accompanied by pieces of dark or crusty French bread. Serves about 8.

Italian Vegetable Antipasto

On individual small plates or one large platter arrange attractively a selection of vegetables that includes some raw ones (radishes, tomatoes, celery, fennel, scallions, mushrooms, and/or green peppers), some pickled ones (mushrooms, artichokes, eggplant, and/or zucchini), and some cold cooked ones (white or red beans, baby carrots, and/or green peas, sprinkled with oil, vinegar, and herbs). Garnish the vegetables with green and/or black olives and cheese cubes, if desired. Serve as an appetizer or as a first course for luncheon or dinner.

Onion-Olive Pizza from Provence

This French version of pizza, called *pissaladiera*, is a specialty of Nice, a lovely port city in Provence. It features the humble onion, long considered indispensable to basic cooking. Chives, garlic, leeks, scallions, shallots, and onions in three colors—white, red, and yellow—are all members of the lily family. Red ones have a sweet flavor; white ones are mild; yellow onions have the strongest flavor.

8 medium-sized onions (about 6 cups), peeled and sliced	1 packaged pizza dough mix (6¾ ounces) or equivalent homemade yeast dough
2 garlic cloves, crushed	1 can (2 ounces) flat anchovies
½ cup olive or vegetable oil	12 pitted black olives
2 tablespoons butter or margarine	
Salt and freshly ground pepper to taste	

Sauté onions and garlic in heated oil and butter in a skillet until soft, being careful not to brown. Remove from stove and add salt and pepper. Cool. Prepare and roll out dough. Arrange in a 12-inch pizza pan or round cookie sheet. Spread onions evenly over the top. Make a lattice pattern with anchovies over onions. Place an olive in center of each square. Brush lightly with oil. Bake in preheated 425° F. oven until dough is golden and crusty, about 20 minutes. Cool slightly. To serve, cut into pie-shaped wedges. Serves 10 to 12.

Balkan Raw Vegetable-Yogurt Dip

Weight watchers will enjoy this flavorful combination made with colorful nutritious vegetables and plain yogurt, once called the milk of eternal life, but actually a healthful cultured milk, low in butterfat and calories. Yogurt has long been a staple food in Balkan countries.

2 cups plain yogurt	½ cup chopped green peppers
½ cup minced green onions, with some tops	1 to 2 teaspoons minced green or red chili peppers
½ cup chopped peeled cucumbers	Salt and pepper to taste
½ cup chopped sweet red peppers	

Combine all ingredients in a bowl. Chill. Serve with pieces of dark bread or sesame crackers. Serves 4 to 6.

Sicilian Caponata

This southern Italian sweet-sour appetizer, made with a fascinating medley of flavorful vegetables and seasonings, can be prepared and kept in a tightly closed jar in the refrigerator for several days to be used as needed.

- 1 medium-sized eggplant, about 1¼ pounds, washed
- Salt
- About ½ cup olive or vegetable oil
- 1 large onion, peeled and chopped
- 2 medium stalks celery, cleaned and chopped
- 2 drained canned Italian plum tomatoes, chopped
- 2 tablespoons tomato paste
- 2 anchovy fillets, rinsed and minced
- 2 tablespoons drained capers
- 2 tablespoons chopped black olives
- 1½ teaspoons sugar
- Freshly ground pepper
- 3 to 4 tablespoons wine vinegar
- ¼ cup chopped fresh parsley

Cut unpeeled eggplant into 1-inch cubes. Put in a colander and sprinkle with salt. Leave to drain for about half an hour. Pat dry with paper toweling and set aside. Heat ⅓ cup oil in a large skillet and add onion and celery. Sauté until tender. Push aside and add eggplant cubes, several at a time. Cook over brisk heat, stirring almost constantly, until soft. Add more oil, as needed. Stir in remaining ingredients, except parsley, and cook slowly, uncovered, until vegetables are tender and flavors well combined, about 15 minutes. Remove from heat. Mix in parsley. Refrigerate at least 24 hours before serving to blend flavors. Serve with chunks of crusty French or Italian bread. Makes about 3 cups.

Vegetable Tempura from Japan

Japanese *tempura*—vegetables dipped in a light thin batter and deep-fried—is an excellent appetizer for any occasion but is particularly appealing at an outdoor party with each guest acting as his own cook.

The vegetables, batter, and sauce can be prepared beforehand and then cooked and served as each batch is ready—crisp and tender. A table can be set with Japanese dishes and the *tempura* eaten with chopsticks, if desired.

The vegetables should be very fresh and can be washed, dried, and cut beforehand if kept covered in the refrigerator. Use any number of the following vegetables. Sauce and batter are for about one pound of vegetables, which will serve about four people.

Asparagus, cut in 2-inch slices
Carrots, cut in thin, diagonal slices
Celery, cut in thin, diagonal slices
Eggplant, cut in ¼-inch slices
Green beans, parboiled and cut into 2-inch slices
Scallions, cut in 1½- or 2-inch lengths
Green pepper, cut in strips
Mushrooms, left whole if small, or cut into halves or quarters if large
Onions, peeled and cut into rings
Sweet potatoes, peeled and sliced ⅛ inch thick
Zucchini, cut into ¼-inch slices

Sauce: Combine 1½ cups *dashi*—a fish stock—or consommé; ⅓ cup soy sauce; ⅓ cup Japanese sweet wine (*mirin*) or dry sherry; and pepper to taste, in a saucepan and bring to a boil. Serve hot in small bowls. Accompany with grated white radish and grated fresh ginger root, if desired.

Batter: Beat 1 large egg with ⅔ cup water in a bowl. Slowly stir in ⅔ cup all-purpose flour and dash of salt until blended. Do not beat. The batter should be lumpy.

To Cook: Heat any fresh flavorless oil, such as sesame, peanut, or corn, in a *tempura nabe* (cooker), electric skillet, or deep fryer until 350° or 355° F. Dip prepared and dried vegetables, a few at a time, into batter and drop at once into hot fat. Turn with chopsticks or a large fork after a minute or two and cook until golden and crisp. Serve at once. Dip in sauce before eating.

Savory Soups

Of all the world's superb soups, those made with vegetables are particularly notable for their rich and interesting variety.

Since the beginning of cookery, man has combined one or more vegetables and the necessary liquid, with or without other foods, in a pot to provide daily sustenance. Every cuisine now has several basic vegetable soups, ranging from the robust and hearty to more aristocratic combinations that are light and designed to stimulate the appetite.

Homemade soups are not only healthful and economical but quite wonderful in flavor. The so-called family soups have much in common; usually thick, they stick to the ribs and are often eaten as a complete meal. The best of them all is still old-fashioned vegetable soup, made with a number of fresh and dried vegetables—often whatever varieties the cook happens to have on hand. Sometimes meat and bones are included, but the cookery is always slow and lengthy.

Particularly desirable soup vegetables are any of the onion family, especially leeks and yellow onions; carrots; white turnips; green cabbage; potatoes; and dried beans. But in the various countries of the world, just about every kind of vegetable has been utilized in making soup. Also very important are herbs, either fresh or dried, which are often made into a *bouquet garni*, three or more herbs—generally bay leaf, thyme, and parsley—tied in a cheesecloth bag, which is removed and discarded before the soup is served.

Old-fashioned methods of preparing some vegetable soups involved a great deal of work and hours of cooking stocks and broths. Although the flavor of a good homemade stock or broth cannot be equaled, cooks today are fortunate in having ready-made consommés, bouillon cubes, and canned broths, which are excellent for soup making.

A vegetable soup is superb for every occasion. Elegant, light ones make appealing first courses. More substantial creations may be served as one-dish meals. Soups are splendid for brunches, luncheons, suppers, or late-evening parties. Orientals enjoy having soup for breakfast, as a between-meal snack, and to refresh the palate after dining. Europeans and Middle Easterners drink soup in the wee hours of the morning as a restorative after a night of revelry.

There are so many fine vegetable soups in the world that it is not possible to do justice to them all. This collection, however, presents a rewarding repertoire of classic recipes, as well as some innovative ones, for family and company meals.

Irish Leek and Potato Soup

Leeks were treasured vegetables in Ireland long before the arrival of potatoes from the New World but eventually both became favorite staples and were combined together in a number of delicious and appetizing dishes. The Irish sometimes call the edible white parts of leeks "swans' necks."

- 4 leeks, white parts only, trimmed, washed, and sliced
- 2 tablespoons butter or margarine
- 4 medium-sized potatoes, peeled and sliced
- 2 cups water
- 1 bouquet garni (1 bay leaf, 2 parsley sprigs, ¼ teaspoon dried thyme)
- Salt and pepper to taste
- 2 cups milk
- 1 cup light or heavy cream
- 2 tablespoons chopped chives or parsley

Sauté leeks in heated butter in a large saucepan until tender. Add potatoes, water, *bouquet garni*, salt and pepper, and bring just to a boil. Lower heat and cook slowly, covered, until potatoes are tender, about 25 minutes. Remove from heat and discard *bouquet garni*. Purée or whirl in a blender. Return to heat and gradually add milk and then cream. Gently reheat but do not boil. Serve garnished with chives or parsley. Serves 4 to 6.

Note: This soup can also be made with yellow onions, instead of leeks, if desired.

Winter Vegetable Soup from Russia

A hearty nourishing soup for a family supper. Serve with pumpernickel or rye bread.

1 large onion, peeled and sliced thinly	3 tablespoons bacon fat or shortening
1 leek, white part only, trimmed, washed, and sliced	8 cups beef bouillon
2 medium-sized carrots, scraped and sliced	1 medium-sized head green cabbage, about 1½ pounds, cored, washed, and shredded
1 stalk celery, trimmed and sliced	1 can (6 ounces) tomato paste
½ white turnip, peeled and cubed	Salt and pepper to taste

Sauté onion, leek, carrots, celery, and turnip in heated fat or shortening in a large kettle for 5 minutes. Add bouillon and bring to a boil. Stir in cabbage and tomato paste, and reduce heat. Season with salt and pepper. Cook slowly, covered, for about 1 hour, or until vegetables are done and flavors have blended. Serves 6 to 8.

German Asparagus Suppe

In Germany this soup is traditionally made with the highly prized white asparagus called *spargel*, but green asparagus is a good substitute. Serve as a first course at luncheon or dinner.

- 1 package (10 ounces) frozen asparagus spears
- 3 tablespoons minced onion
- 2 cups chicken broth
- Salt and pepper to taste
- 2 tablespoons butter or margarine
- 2 tablespoons all-purpose flour
- 2 cups milk
- 1 cup light or heavy cream
- 1/8 teaspoon freshly grated or ground nutmeg

Cook asparagus according to package directions until tender. Drain, reserving 1/2 cup liquid. Cut asparagus into pieces, reserving the tips. Combine pieces, reserved liquid, onion, and chicken broth in a saucepan, and cook over medium heat for 5 minutes. Purée or whirl in a blender. Season with salt and pepper. Melt butter in a medium saucepan. Stir in flour to form a *roux*. Cook 1 minute. Gradually add milk and cook slowly, stirring, until thick and smooth. Mix in asparagus purée and cream. Leave on stove long enough to heat through. Add nutmeg and serve garnished with asparagus tips. Serves 4 to 6.

Pumpkin Soup from Venezuela

A gourdlike fruit native to America, the pumpkin is used in Central and South American countries to make imaginative dishes, such as this soup. The sweet-flavored orange flesh has a generous supply of vitamin A.

- 2 tablespoons butter or margarine
- 2 tablespoons all-purpose flour
- 2 1/2 cups milk
- 3 to 4 tablespoons brown sugar
- 1/4 teaspoon ground nutmeg
- 1/4 teaspoon ground cinnamon
- 1 teaspoon salt
- 2 cups mashed cooked or canned pumpkin
- 1/2 cup beef consommé
- 2 egg yolks

Melt butter in a saucepan and stir in flour. Cook 1 minute. Gradually add milk and cook slowly, stirring, until thick and smooth. Add remaining ingredients, except egg yolks, and simmer 15 minutes. Spoon some of hot mixture into egg yolks. Return to soup. Simmer 1 or 2 minutes longer. Serves 6.

Czech Cream of Cauliflower Soup

The cauliflower—actually a variety of cabbage—was developed for its unique head of florets, which are usually white but can be green or even purple in color. It is a delicate, nutritious vegetable, particularly prized by Czech cooks who use it in a number of delectable dishes.

1 medium-sized head cauliflower	2 tablespoons all-purpose flour
2 cups water	1 bay leaf
1 teaspoon salt	1/2 teaspoon dried thyme
2 tablespoons fresh lemon juice	Salt and pepper to taste
2 tablespoons butter or margarine	3 cups light cream or milk
	2 tablespoons chopped fresh parsley

Wash cauliflower and cut off stem and tough outer leaves. Put in a kettle with water, salt, and lemon juice. Cook, uncovered, 5 minutes. Cover and cook another 15 minutes. Drain, reserving liquid. Cut cauliflower into small pieces. Melt butter in a large kettle and stir in flour. Cook 1 minute. Gradually add cauliflower liquid and cook slowly, stirring, until thickened and smooth. Add bay leaf, thyme, salt, pepper, and cream or milk. Cook slowly, stirring, until smooth and thick, about 10 minutes. Add cauliflower and parsley, and cook another 5 minutes. Remove and discard bay leaf. Serves 6 to 8.

Potage Crème de Champignons

An elegant French cream of mushroom soup for a luncheon or dinner party.

1 pound fresh mushrooms	Salt and pepper to taste
¼ cup butter or margarine	5 cups hot light cream or milk
¼ cup finely chopped onion	2 tablespoons chopped fresh dill or parsley
2 tablespoons all-purpose flour	
2 cups chicken broth	
½ teaspoon crumbled dried rosemary	

Clean mushrooms by rinsing quickly or wiping with paper toweling to remove dirt. Cut off any woody stem ends and wipe dry. Slice from round sides through stems. Set aside. Melt butter in a large saucepan and sauté onion in it until tender. Add mushrooms and sauté 4 minutes. Stir in flour. Gradually add chicken broth, stirring constantly, and then rosemary, salt and pepper. Cook slowly, covered, 20 minutes. Add cream and leave on the stove long enough to heat through. Add dill. Serves 6 to 8.

Economic Cream of Lettuce Soup

An inexpensive easy-to-prepare soup with an interesting flavor that derives primarily from lettuce, which is most often served uncooked in salads but is also excellent when cooked.

6 scallions, with tops, cleaned and chopped	3 cups finely chopped leafy lettuce
3 tablespoons butter or margarine	4 cups milk
	Salt and pepper to taste
3 medium-sized stalks celery, with some leaves, cleaned and chopped	3 tablespoons chopped fresh parsley

Sauté scallions in heated butter in a large saucepan until tender. Add celery and lettuce and cook until wilted and tender. Pour in milk. Add salt and pepper. Slowly bring to boiling point and remove from heat. Add parsley. Serves 4 to 6.

Portuguese Kale-Bean Soup

In Portugal this soup is made of a deep-green native kale, which is unlike any grown elsewhere. American kale, a variety with large curly leaves, a robust flavor, and a high vitamin content, can be used as a substitute. *Chorizo* is a highly spiced sausage relished throughout the Iberian peninsula.

- 1 cup dried pea or navy beans
- 6 cups water
- 1 teaspoon salt
- 1 package (10 ounces) fresh kale
- 1 large onion, peeled and chopped
- 2 tablespoons olive or vegetable oil
- 2 cups cubed potatoes
- 1/2 pound chorizo sausages, sliced (optional)
- 1 tablespoon vinegar
- Salt and pepper to taste

Put beans, water, and salt in a large kettle and bring to a boil. Boil 2 minutes. Remove from heat and let stand, covered, for 1 hour. Wash kale and cut off tough stems and midribs. Tear large leaves into small pieces. Sauté onion in heated oil in a large kettle until tender. Add beans with liquid and bring to a boil. Lower heat and cook slowly 45 minutes. Add kale, potatoes, and *chorizo*, and cook slowly, covered, about another hour, or until ingredients are tender. Mix in vinegar and season with salt and pepper. Serves 6 to 8.

Iranian Cold Cucumber-Yogurt Soup

Iranians prepare this popular summer soup with exotic embellishments such as nuts, raisins, and fresh herbs. Customarily the soup includes ice cubes, but these are not necessary if it is well chilled.

- 2 medium-sized cucumbers, peeled, seeded, and chopped (about 2 cups)
- Salt
- 2 cups plain yogurt
- 1/4 cup chopped scallions
- 1/2 cup golden raisins
- 1/4 cup chopped walnuts
- 1/4 cup chopped fresh mint or dill

Put cucumbers in a strainer or colander and sprinkle with salt. Leave to drain for 30 minutes. Combine with remaining ingredients, except mint, and chill. Serve garnished with mint. Serves 4.

Note: For a thinner soup add a little water.

Gazpacho from Andalusia

Sometimes termed a soup-salad, gazpacho, which originated in Spain's southern province of Andalusia, has become a great summer favorite in America. This is the traditional recipe.

1 or 2 garlic cloves	3 tablespoons olive or vegetable oil
1 teaspoon salt	2 cups cold water
1 cup soft bread cubes, crusts removed	2 cups tomato juice
¼ cup wine vinegar	Pepper to taste
4 medium-sized ripe tomatoes, peeled and diced	Garnishes: About ¾ cup chopped scallions and 1 cup toasted bread cubes
1 medium-sized cucumber, peeled and diced	
1 large green pepper, cleaned and diced	

Pound garlic and salt together in a mortar with a pestle or mash in a bowl with a wooden spoon. Add bread cubes and vinegar, and work into a paste. Add 3 tomatoes and ½ cucumber and mash as fine as possible (or whirl in a blender). Mix with remaining diced tomato, cucumber, and green pepper. Chill. Just before serving, stir in oil, water, and tomato juice. Season with salt and pepper, and serve topped with garnishes. Serves 6 to 8.

Note: For a tarter flavor, add more vinegar.

Hong Kong Spinach-Mushroom Soup

2 tablespoons sliced scallions with tops	1/4 pound fresh mushrooms, cleaned and thinly sliced
1 tablespoon peanut or vegetable oil	4 ounces fine noodles, cooked and drained
1/2 cup sliced bamboo shoots	Salt and pepper to taste
6 cups chicken broth	
1/2 pound fresh spinach, washed and chopped	

Sauté scallions until tender in heated oil in a large saucepan. Add bamboo shoots and broth and bring to a boil. Stir in spinach and mushrooms and cook, covered, over medium heat about 12 minutes, until spinach is tender. Add egg noodles, salt and pepper, and cook another 2 or 3 minutes. Serves 6 to 8.

Cold Broccoli Soup

Broccoli, a dark-green vegetable related to cauliflower, also has odd small heads, called curds, and a thick stem that are both edible. Broccoli, very rich in vitamin C, should be cooked until just tender and still crisp. It marries well with sour cream and vinegar or lemon juice.

1 medium-sized onion, peeled and minced	cups chopped cooked fresh broccoli
2 tablespoons butter or margarine	1/2 teaspoon dried basil
4 cups chicken broth	Salt and pepper to taste
2 packages (10 ounces each) frozen chopped broccoli or 4	3 tablespoons all-purpose flour
	1 1/2 cups light cream or milk

Sauté onion in heated butter in a saucepan until tender. Add broth and bring to a boil. Mix in broccoli, basil, salt and pepper. Cook slowly, covered, about 12 minutes, or until tender. Stir a few spoonfuls of hot liquid with flour to make a smooth paste. Mix into soup. Bring to a boil. Remove from stove and strain or whirl in a blender. Combine with cream or milk and chill. Serves 8.

Soupe au Pistou

This favorite soup of France's lovely Provence is very similar to minestrone, but it includes a spicy sauce, or *pistou*, made of crushed garlic, olive oil, grated cheese, and fresh basil, which is added at the end of the cooking. The French relish the soup in early spring when it is prepared with small fresh white beans. Dried white navy beans are a good substitute.

3	tablespoons butter or olive oil	1	can (1 pound) cannellini or navy beans, drained; or 1½ cups cooked dried white beans, pea or navy
1	large onion, peeled and diced		
2	leeks, white parts only, trimmed, washed and sliced (optional)		
		½	cup broken spaghettini or vermicelli
2	large tomatoes, peeled and chopped	3	garlic cloves, crushed or minced
12	cups water		
	Salt and pepper to taste	½	cup chopped fresh basil, or 1½ tablespoons dried basil
2	cups diced raw potatoes		
2	cups cut-up green beans	½	cup freshly grated Parmesan cheese
2	unpeeled medium-sized zucchini, washed and diced		
		¼	cup olive oil

Heat butter or oil in a large kettle. Add onion and leeks and sauté over low heat until tender. Add tomatoes and cook about 3 minutes, until mushy. Add water and bring to a boil. Season with salt and pepper. Mix in potatoes and green beans. Lower heat and simmer, uncovered, for 15 minutes. Add zucchini, cannellini or beans, and spaghettini and cook another 15 minutes or until vegetables are tender. While the soup is cooking, prepare the *pistou* sauce. Pound garlic and basil together to form a paste in a mortar with a pestle or mash them in a bowl with a wooden spoon. Stir in cheese. Add oil, 1 tablespoon at a time, and beat to make a thick paste. Just before serving, add 2 cups hot soup to the paste. Slowly stir into hot soup and serve at once. Pass grated cheese with soup. Serves 12.

Note: Pistou sauce can be purchased in cans in some specialty food stores.

Spring Soup

A nourishing, elegant soup made with an assortment of greens.

- 2 tablespoons butter or margarine
- 6 scallions, with tops, washed and minced
- ½ cup shredded leafy lettuce
- 1 cup shredded fresh spinach
- ½ cup chopped watercress
- ½ teaspoon dried chervil or parsley
- 2 teaspoons sugar
- 6 cups chicken broth
- Salt and pepper to taste
- 1 cup light cream or milk

Melt butter in a large saucepan and sauté scallions in it until tender. Add lettuce, spinach, watercress, and chervil, or parsley. Simmer, covered, for 10 minutes. Add sugar, broth, salt and pepper, and simmer, covered, 30 minutes. Add milk and leave on stove long enough to heat through. Serves 6.

Potage Crécy

There are many versions of this French carrot soup, named for a town in France that is famous for its golden carrots. This recipe is one of the best.

- 1 pound, about 8 medium-sized, carrots
- 3 tablespoons butter or margarine
- 1 medium-sized onion, peeled and chopped
- 2 medium-sized potatoes, peeled and chopped
- 1 bouquet garni (1 bay leaf, 2 parsley sprigs, ¼ teaspoon dried thyme)
- ½ teaspoon celery seed
- Salt and pepper to taste
- 3 cups water
- 2 cups light cream or milk
- ¼ cup chopped fresh parsley

Wash and scrape carrots; cut into thin slices. Melt 2 tablespoons butter in a large saucepan and sauté onion until tender. Add carrots, potatoes, *bouquet garni,* celery seed, salt and pepper, and water. Bring to a boil. Lower heat and cook slowly, covered, until vegetables are tender, about 25 minutes. Remove and discard *bouquet*

SAVORY SOUPS

garni. Purée or whirl in a blender. Return to saucepan and reheat. Add cream and cook slowly for 5 minutes. Mix in remaining 1 tablespoon butter and parsley. Serves 6 to 8.

Peruvian Mixed Vegetable Soup

1 large onion, peeled and chopped	¼ teaspoon dried thyme
2 garlic cloves, crushed	Salt and pepper to taste
2 tablespoons butter or margarine	4 cups water
1 can (1 pound) tomatoes, undrained	1 cup frozen green peas
½ cup uncooked rice	1 cup canned chick-peas, drained
	3 tablespoons chopped fresh parsley

Sauté onion and garlic in heated butter in a large saucepan until tender. Add tomatoes, rice, thyme, salt and pepper, and water. Cook slowly, covered, for 25 minutes. Add green peas and chick-peas and cook about 10 minutes longer, until ingredients are tender. Add parsley. Serves 6.

Soupe à l'Oignon Gratinée

This famous traditional French onion soup is a good supper dish, but it is also a superb late-evening snack, especially during the winter.

1½ pounds (about 5 cups sliced) onions, peeled and thinly sliced	Salt and pepper to taste
About 5 tablespoons butter or margarine	6 cups beef bouillon
3 tablespoons vegetable oil	½ cup dry white wine (optional)
1 teaspoon sugar	6 slices toasted French bread
	About 1½ cups grated Gruyère or Swiss cheese

Sauté onions in 3 tablespoons heated butter and the oil in a large heavy saucepan over moderate heat until tender and translucent, being careful not to brown. Add sugar and mix well. Season with salt and pepper. Pour in bouillon and bring to a boil. Lower heat and simmer, covered, for 20 minutes. Add wine and continue to cook slowly for 10 minutes. Ladle into earthenware or other ovenproof bowls. Top each with a slice of toasted French bread. Sprinkle generously with cheese and a little melted butter. Put in a preheated 375° F. oven for 20 minutes, or until cheese is melted. Then slide under heated broiler for a minute or two, until golden and crusty on top. Serve in bowls. Serves 6.

Circassian Vegetable-Yogurt Soup

This soup is named for the lovely Circassian ladies who were brought in yesteryear from southern Russia to the Sultan's court at Constantinople. A number of dishes still eaten in the Middle East are credited to them.

1	beef soup bone with meat, about 1¼ pounds	1	cup diced carrots
1½	teaspoons salt	1	cup cut-up green beans
½	teaspoon pepper	1	cup green peas
1	medium-sized bay leaf	2	cups tomato juice
6	cups water	½	teaspoon dried oregano or thyme
1	large onion, peeled and chopped		Dash cayenne
⅓	cup uncooked rice	1	cup plain yogurt or sour cream at room temperature

Put soup bone, salt, pepper, bay leaf, and water in a large heavy kettle. Bring to a boil and skim. Lower heat and cook slowly, covered, for 1½ hours. Skim again. Add onion, rice, carrots, beans, peas, tomato juice, oregano, and cayenne. Continue to cook slowly, covered, for about 30 minutes, or until meat, rice, and vegetables are tender. Remove and discard bay leaf. Take out soup bone, and cut any meat into small pieces. Discard bone and return meat to soup. Stir in yogurt or sour cream and leave on stove long enough to heat through. Serves 10 to 12.

Tuscan-Bean Pasta Soup

In the northern Italian province of Tuscany many of the best dishes are made with the native fat and tender white beans, which are eaten in such quantity that Tuscans are mockingly called "bean eaters." Any large white bean may be used as a substitute.

1 cup dried white beans	2 large tomatoes, peeled and chopped
12 cups water	Salt and pepper to taste
1 cup diced ham, including some fat	1 cup elbow macaroni or broken spaghetti
2 tablespoons olive or vegetable oil	1/4 cup chopped fresh parsley
1 large onion, peeled and diced	About 1 cup grated Parmesan cheese
1 or 2 garlic cloves, crushed	
2 stalks celery, with leaves, chopped	

Rinse and pick over beans; then put them in a large saucepan and cover them with water. Bring to a boil; boil 2 minutes. Remove from heat and leave, covered, 1 hour. Put ham, oil, onion, and garlic in a kettle and sauté, stirring, for 5 minutes. Add beans and liquid, celery, tomatoes, salt and pepper, and bring to a boil. Lower heat and cook, covered, about 1½ hours, or until beans are tender. Uncover and bring soup to a boil. Add macaroni and cook until tender about 10 minutes. Stir in parsley. Serve with grated Parmesan cheese. Serves 8 to 10.

Down East Corn-Potato Chowder

Early colonists in Maine utilized native American foods, such as corn and potatoes to make nourishing soups that became known as chowders, their name being derived from the utensil in which such soups were cooked: a large French kettle called *la chaudière*, which was brought to the northern New England coast by French fishermen.

¼ cup diced salt pork	2 cups whole corn kernels, fresh, frozen or canned
1 medium-sized onion, peeled and sliced	1½ cups milk
2 cups boiling water	Salt and pepper to taste
2 cups diced peeled raw potatoes	4 common crackers, split in halves (optional)

Fry salt pork to release all fat in a large saucepan or heavy kettle. Add onion and sauté until tender. Add water and potatoes, and cook until potatoes are just tender, about 12 minutes. Stir in corn and cook about 5 minutes longer, cooking a little longer for fresh corn than the other kinds. Add milk, season with salt and pepper, and leave on the stove long enough to heat through. Serve hot in soup plates ladled over crackers. Serves 4.

Russian Cold Beet Soup with Sour Cream

This is an easy-to-prepare version of the famous *borscht* that Russian cooks prepare in many flavorful variations. Beets, grown in several varieties and colors, are nutritious, and should always be cooked unpeeled with some of the leaf stem attached.

2 cups or 1-pound can shoestring beets or equivalent amount whole beets cut in julienne strips	3 tablespoons cider vinegar
	1 teaspoon sugar
	Salt and pepper to taste
	½ cup sour cream
4 cups beef bouillon	About 2 tablespoons chopped fresh dill or parsley
1 large bay leaf	

Combine beets, bouillon, bay leaf, vinegar, sugar, salt and pepper, in a saucepan. Bring to a boil. Lower heat and cook slowly, covered, 10 minutes. Chill. Serve in soup bowls, each garnished with a spoonful of sour cream and some dill. Serves 6.

Mexican Corn-and-Chili Caldo

In Mexico a *caldo* is generally a simple clear broth, but it may also include vegetables. Serve with warm corn bread or muffins.

1	large onion, peeled and chopped	½	teaspoon dried oregano
			Salt and pepper to taste
1	or 2 garlic cloves, crushed	1½	cups fresh or frozen corn niblets
2	tablespoons vegetable oil		
1	to 2 tablespoons chili powder	½	cup fresh or frozen cut-up green beans
2	medium-sized tomatoes, peeled and chopped		
		3	cups milk

Sauté onion and garlic in heated oil in a large saucepan until tender. Add chili powder and cook 1 minute. Stir in tomatoes, oregano, salt and pepper. Mix in corn, green beans, and milk and cook over low heat, for about 15 minutes, until ingredients are tender. Allow less time for frozen vegetables. Serves 4 to 6.

Finnish Summer Vegetable Soup

An excellent colorful soup that is best made when fresh vegetables are at their prime during the summer, but it can also be prepared with frozen vegetables.

2	cups thinly sliced peeled onions	1	tablespoon sugar
			Salt and pepper to taste
1	cup thinly sliced scraped carrots	6	tablespoons all-purpose flour
		6	cups hot milk
2	cups cut-up cauliflower	2	tablespoons butter or margarine
2	cups cut-up green beans		
1	cup green peas	⅓	cup chopped fresh parsley
5	cups boiling water		

Put vegetables and boiling water in a large kettle. Add sugar, salt and pepper, and bring to a boil. Lower heat and cook slowly, covered, until vegetables are just tender, about 25 minutes. Combine flour and milk, and mix until smooth. Gradually stir into soup and mix well. Continue to cook several minutes longer, until liquid has thickened and vegetables are tender. Remove from heat. Add butter and parsley. Serves 8 to 10.

Creole Tomato-Rice Gumbo

A gumbo is a thick well-seasoned Creole soup that takes its name from the Bantu word for okra, a main ingredient that has a mucilaginous quality and is both a thickening agent and vegetable. Okra is often called gumbo because of its popular inclusion in the soup, but the green pods can also be breaded and fried or cooked with other vegetables and added to stews and salads.

- 1 medium-sized onion, peeled and chopped
- 1/2 cup diced green pepper
- 1/2 cup diced celery
- 3 tablespoons butter or margarine
- 3 tablespoons all-purpose flour
- 3 cups chicken broth
- 1 can (1 pound) tomatoes, undrained
- 1 small bay leaf
- 1/2 teaspoon dried thyme
- Salt and pepper to taste
- 1/3 cup uncooked long-grain rice
- 1 cup fresh or frozen cut-up okra
- Dash of Tabasco or hot pepper sauce

Sauté onion, green pepper, and celery in heated butter in a large saucepan for 5 minutes. Stir in flour. Gradually add chicken broth, stirring constantly while adding. Mix in tomatoes and break up with a fork. Add bay leaf, thyme, salt, pepper, and rice and cook slowly, covered, about 30 minutes. Stir in okra and Tabasco or hot pepper sauce, and cook about 10 minutes longer; allow a little less time for frozen okra. Remove and discard bay leaf. Serves 6 to 8.

Minestrone alla Romana

This well-known, hearty, thick vegetable soup is made in several variations but usually includes a number of dried and fresh vegetables and either pasta or rice. The name is taken from the Latin word *minister*—to serve, dish up—because long ago monks kept pots of the soup on the stove to give to weary and hungry travelers. This version comes from Rome.

3	thin slices bacon, chopped	8	cups beef bouillon or water
1	tablespoon olive or vegetable oil		Salt and pepper to taste
1	large onion, peeled and chopped	1	can (1 pound) white beans, drained; or 1½ cups cooked dried white beans, pea or navy
1	or 2 garlic cloves, crushed		
1	cup chopped scraped carrots	1	cup small pasta or broken spaghetti
2	cups chopped green cabbage		
2	unpeeled zucchini, stems trimmed, sliced		About ½ cup grated Parmesan cheese, preferably freshly grated
1	can (1 pound) tomatoes, undrained		
1½	cups diced peeled raw potatoes		

Combine bacon, oil, onion, and garlic in a large kettle and sauté over low heat for 5 minutes. Add carrots and cabbage, and sauté another 5 minutes. Add zucchini, tomatoes, potatoes, and bouillon, or water, and bring to a boil. Season with salt and pepper. Lower heat and cook slowly, covered, until ingredients are just tender, about 30 minutes. Add beans and pasta, and cook over moderate heat until pasta is tender, about 12 minutes longer. Serve with grated Parmesan cheese. Serves 8 to 10.

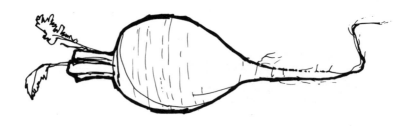

Stuffed Vegetables

Over the years ingenious cooks have created a fascinating variety of delectable dishes by stuffing vegetables, or sometimes only their leaves, with savory mixtures. The diverse flavors and textures of vegetable and stuffing serve to complement each other.

Excellent when served as an appetizer, as a main course for luncheon or supper, or as an accompaniment to cold meat or poultry, these vegetable dishes are particularly recommended for buffets. Easily prepared beforehand, either hot or cold, they provide the hostess with an interesting variation in vegetable cookery.

Very important to the success of these dishes is the careful selection of only the best vegetables, choosing those that are fresh, well colored, and firm. In some cases they should also be of uniform size and shape.

In preparing the vegetables, remove the insides with care so the shells are not broken or cut. For some vegetables, such as green peppers, onions, squash, and cabbage leaves, it is preferable that they first be blanched. For this, use as little liquid as possible, so the nutrients will be preserved. After blanching, drain thoroughly.

Once the vegetables are stuffed, they should be cooked, partially covered with liquid, on top of the stove or, with fats or liquids, in the oven. Either way they should not be overcooked as they will lose their flavor and shape. (Of course some vegetables, such as mushrooms, tomatoes, and green peppers, may be stuffed and left uncooked, depending on the mixture.) Since vegetables can vary considerably in size, the precise cooking time in each case is difficult to predict. Consequently it is preferable to check the vegetables at least once or twice during cooking to be sure they will not be overdone.

There are many possibilities for the stuffing of vegetables—bread crumbs, cheese, rice, and other grains; meats, poultry, and seafood—often with flavorings ranging from herbs and spices to nuts and raisins. Sometimes, due to variations in size, the exact amount of stuffing is not easy to specify. Any leftover stuffing can be formed into balls and cooked in broth or sautéed in oil or butter.

Since the stuffed vegetables are in themselves colorful and attractive, garnishes are generally not necessary, but a delightful way to present them is to use serving dishes that have a vegetable shape or motif.

Many of the recipes that follow are from Mediterranean countries where this type of cookery began in ancient times and is still a specialty; there are also recipes from other foreign lands.

Corsican Stuffed Onions

8 medium-sized yellow onions
Boiling salted water
1 pound sausage meat
2 slices stale bread
2 egg yolks, beaten
1 tablespoon chopped capers
3 tablespoons grated Parmesan cheese
⅓ cup chopped fresh herbs
(parsley, thyme, rosemary, oregano, basil)
Salt and freshly ground pepper to taste
About 4 tablespoons fine dry bread crumbs
About 2 tablespoons olive or vegetable oil
8 parsley sprigs

Cook onions, unpeeled, in a saucepan of boiling salted water for 15 minutes. Drain, rinse in cold water, and peel. Cut a slice from top of each onion and carefully take out the center, leaving an outer ring of 4 layers. Discard the onion centers or reserve for another dish.

Cook sausage, separating with a fork, until redness disappears. Remove from heat and pour off all except 1 tablespoon of fat. Soak slices of bread in water until soft and squeeze dry. Add to sausage along with egg yolks, capers, cheese, and herbs. Season with pepper. Cook, stirring constantly, for 2 minutes. Mix well and remove from heat. Spoon into onion shells, packing well and shaping into a mound at the top. Sprinkle tops with bread crumbs and oil. Arrange in a greased shallow baking dish. Bake, covered, in preheated 375° F. oven until onions are tender, about 30 minutes. Remove from oven and garnish each onion with a sprig of parsley. Serves 8.

Florida Tuna-Stuffed Green Peppers

Our native American capsicum peppers are grown in hundreds of varieties. Large sweet ones can be colored green, red, or yellow. The bell pepper changes its color as it ages on the vine. Green peppers are picked when unripe, and red peppers are those left to mature. Both have just about the same mild flavor and are excellent for stuffing.

4 medium-sized green peppers	1 teaspoon Worcestershire sauce
1/4 cup minced onion	1/2 teaspoon dried basil
2 tablespoons vegetable oil or butter	Salt and pepper to taste
2 tablespoons all-purpose flour	1 can (about 7 ounces) tuna fish, drained and flaked
1 cup tomato juice	About 1/4 cup grated Parmesan cheese
1 teaspoon fresh lemon juice	

Wash peppers and slice off tops. Scoop out and discard seeds and fibers. Rinse pepper shells. Blanch in boiling salted water for 5 minutes and drain well. Meanwhile, sauté onion in heated oil in a small saucepan until tender. Stir in flour and cook 1 minute. Add tomato and lemon juice, and cook slowly, stirring, until thickened. Mix in remaining ingredients, except cheese. Spoon into peppers. Arrange in a greased shallow baking dish. Sprinkle tops with cheese. Bake in a preheated 350° F. oven for 20 minutes. Serves 4.

Mexican Corn-Stuffed Zucchini

A very colorful luncheon entrée or accompaniment to meat at dinner.

- 4 zucchini, about ½ pound each
- 2 cups frozen corn kernels, defrosted, or canned corn
- 1 egg
- 2 tablespoons milk
- 2 medium-sized tomatoes, peeled and chopped
- ½ teaspoon dried basil
- Salt and pepper to taste
- ½ cup grated Parmesan cheese
- About 2 tablespoons butter or margarine

Wash zucchini. Cut off stem ends and split lengthwise. Scoop out pulp to form shells, discarding it or reserving it to cook later as a separate dish. Arrange zucchini shells in a shallow greased baking dish. Combine corn, egg, milk, tomatoes, basil, salt and pepper, and spoon into zucchini shells. Cover with foil and bake in a preheated 350° F. oven for 30 minutes. Remove from heat. Sprinkle with cheese and dot with butter. Bake, uncovered, for 10 minutes. Serves 4.

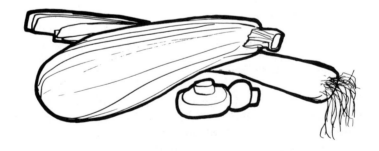

Israeli Stuffed Cabbage

The utilitarian and familiar cabbage is available in a great many varieties, but the heads are usually classified by their leaves—plain or curly—and their color—white, green, or red. Of all the numerous ways of cooking cabbage, one of the most interesting has long been stuffing the leaves with a savory filling. This recipe is one of the best.

- 1 head, about 4 pounds, firm green cabbage
- Boiling salted water
- 1 pound lean ground beef
- 1 small onion, peeled and minced
- ½ cup uncooked rice
- 2 tablespoons grated raw carrot
- 3 tablespoons currants
- 2 eggs, beaten
- Salt and pepper to taste
- 2 cans (8 ounces each) tomato sauce
- ½ cup brown sugar
- ⅓ cup fresh lemon juice

Remove and discard any wilted outer leaves of cabbage. With a sharp knife cut out core to a depth of about 3 inches. Put in a large kettle of boiling salted water and parboil over high heat about 12 minutes. Remove with tongs and take off large outer leaves which have cooked long enough to be tender. Cook remaining cabbage until leaves are soft, another 5 to 10 minutes. Remove from water and take off leaves. Spread out leaves on a wooden board and cut out tough rib from each leaf.

Cook beef in a skillet, stirring often with a fork to separate, until redness disappears. Remove from heat and pour off any fat. Add onion, rice, grated carrot, currants, and eggs. Season with salt and pepper. Form 1 to 2 tablespoons of meat mixture into a small ball and place in the center of each leaf. The exact size of the ball will depend on the size of the leaf. Fold over edges and roll tightly. Arrange, seam sides down, one next to the other, in layers in a heavy casserole. Combine tomato sauce, brown sugar, and lemon juice. Pour over stuffed cabbage leaves. Cook slowly, tightly covered, about 50 minutes, or until cooked. Add a little water during cooking, if necessary. Serves 6 to 8.

French Mushroom-Stuffed Eggplant

In this recipe the elegant eggplant is filled with an appealing French mushroom stuffing called *duxelles,* a classic mixture named after the seventeenth-century French gourmet, Marquis d'Uxelles, whose famous chef, LaVarenne, is credited with its creation.

1 medium-sized eggplant, about 1¼ pounds	About 9 tablespoons olive or vegetable oil
1 pound fresh mushrooms	Salt and freshly ground pepper to taste
¼ cup minced onions	
¼ cup minced shallots	⅓ cup chopped fresh parsley
2 garlic cloves, crushed	½ cup grated Swiss cheese
About 9 tablespoons butter or margarine	¼ cup fine dry bread crumbs

Wash eggplant and remove stem from end. Set eggplant aside. Rinse quickly or wipe mushrooms with wet paper toweling to remove dirt. Cut off any tough stem ends. Chop mushrooms, both caps and stems, cutting as fine as possible. Enclose in cheesecloth and squeeze out all moisture. Set aside.

Heat 5 tablespoons butter and 2 tablespoons oil in a large skillet, and sauté onions, shallots, and garlic. Add chopped mushrooms and cook over a high flame about 1 minute, stirring constantly. Lower heat and cook slowly until all moisture has evaporated and mixture is quite dark. Stir now and then. The cooking will take several minutes. Season with salt and pepper. Stir in parsley and remove from stove.

Cut eggplant in half lengthwise and remove pulp leaving a ¼-inch shell. Chop pulp with a silver or stainless-steel knife and sauté in remaining 7 tablespoons of oil until soft. Add more oil, if necessary. When cooked, stir in mushroom mixture and cheese. Correct seasoning. Spoon into eggplant shells. Sprinkle tops with the bread crumbs and remaining butter. Arrange in a shallow greased baking dish and cook in a preheated 375° F. oven about 45 minutes, or until eggplant is fork tender. Serves 2 as an entrée, or 4 as an accompaniment.

Lebanese Stuffed Squash With Yogurt Sauce

Summer squash that is harvested while still young is tender and has a thin skin, so that the entire vegetable is edible. Of the many varieties of squash, those that are warted and yellow, called straightneck or crookneck, are excellent for stuffing.

1 small onion, peeled and minced	Salt and pepper to taste
2 tablespoons vegetable oil	6 medium-sized tender yellow squash
1 pound ground lamb	2 tablespoons all-purpose flour
¾ cup cooked rice	2 cups plain yogurt at room temperature
¼ cup chopped pine nuts	
¾ teaspoon ground cinnamon	¼ cup chopped fresh parsley
¼ teaspoon ground nutmeg	
2 tablespoons chopped fresh mint	

Sauté onion in heated oil in a skillet until tender. Add lamb and cook, stirring frequently to separate meat, until redness disappears. Pour off any fat. Add rice, pine nuts, cinnamon, nutmeg, and mint. Season with salt and pepper. Mix well and remove from heat.

Wash unpeeled squash and cut off stem ends. Cut each one in half lengthwise. Scoop out pulp to form shells, discarding it or reserving it to cook separately later, and leaving an even border about ¼ inch thick of squash. Spoon lamb-rice mixture into squash shells, heaping into a mound at the top. Arrange in one shallow extra large baking dish or in two smaller ones. Add water, ½ inch in depth. Cook, covered, in a preheated 375° F. oven for 45 minutes. Remove from oven and carefully take baked squash out with a slotted spoon and place on a warm plate. Keep warm. Spoon 2 tablespoons liquid from baking dish into a small saucepan and stir in flour. Add yogurt and leave on stove, stirring constantly, until heated through. Stir in parsley and remove from heat. Serve with stuffed squash. Serves 6.

Olive-Chicken–Filled Pimientos from Morocco

The pimiento, a variety of the bell pepper, is heart-shaped, very sweet, and has soft orange-red flesh. It is sometimes called the Spanish paprika and grows only in tropical climates. Pimientos are rarely found fresh in our stores but are widely available canned.

1 cup minced cooked chicken	1/2 teaspoon paprika
1/3 cup finely chopped black olives	4 tablespoons chopped fresh parsley
1 or 2 garlic cloves, crushed	Salt and pepper to taste
2 tablespoons minced onion	4 well-drained canned pimientos
1/4 teaspoon ground coriander	

Combine all ingredients, except pimientos, and mix well. Arrange pimientos in a shallow baking dish and fill with chicken-olive mixture. Bake in preheated 400° F. oven for 15 minutes. Serves 4.

Turkish Lamb-Rice Stuffed Eggplant

2 medium-sized eggplants, about 1 1/4 pounds each, washed	6 tablespoons tomato paste
About 1/2 cup olive or vegetable oil	1/4 cup uncooked rice
	1 tablespoon chopped fresh dill
2 medium-sized onions, peeled and minced	3 tablespoons chopped fresh parsley
1 pound ground lamb	Salt and pepper to taste
1/2 cup beef bouillon	1 cup plain yogurt at room temperature

Cut eggplants in half lengthwise. Remove stems from ends and scoop out insides, leaving a shell about 1/2 inch thick. Dice pulp and set aside. Heat 1/4 cup oil in a large skillet and sauté outer skins of eggplants until just softened. Remove carefully and arrange, hollow sides up, in a shallow baking dish. Add more oil, if needed.

Add onions to drippings and sauté until tender. Add lamb and cook, stirring frequently with a fork, until lamb has lost its redness. Add bouillon and 3 tablespoons tomato paste and cook 5 minutes. Stir in diced eggplant pulp and cook until eggplant is soft, about 10 minutes. Remove from heat and mix in rice, dill, and parsley. Season with salt and pepper. Fill eggplant shells with stuffing. Bake in a preheated 375° F. oven for 45 minutes, or until stuffing is cooked. Combine remaining 3 tablespoons tomato paste and yogurt, and heat but do not boil. Serve with baked eggplant. Serves 4.

Nebraska Stuffed Turnips

There are several varieties of the root vegetable called turnip. The white kind has a white or purple-white tuber and tops that are usually green, but can be tinged with purple. Both parts are edible. The roots have to be peeled, and the vegetable can then be prepared in a number of appetizing dishes such as this one. Small raw white turnips are good in salad.

8 medium-sized white turnips of uniform size	1 small onion, peeled and minced
Boiling salted water	1 teaspoon Worcestershire sauce
1 cup cooked long-grain rice	2 teaspoons chopped fresh parsley
1 cup minced cooked beef, lamb, or veal	Salt and pepper to taste
1 teaspoon curry or turmeric powder	

Pare turnips and cut a slice from top of each one. Scoop out insides and reserve or discard, leaving a shell about ¼ inch thick. Cook shells and top slices in boiling salted water for 15 minutes. Drain and dry. Meanwhile combine remaining ingredients and spoon into turnip shells. Cover with tops. Arrange in a shallow baking dish and add a little of the cooking water. Bake in preheated 350° F. oven 20 minutes, or until tender. Serves 8.

Note: The scooped-out turnip flesh may be cooked and mashed or creamed. There may be some stuffing left over as the amount required will depend on the turnips' size.

Italian Stuffed Mushrooms

12 extra large mushrooms
1 cup minced cooked ham
2 tablespoons minced black olives
1 tablespoon minced blanched almonds
2 garlic cloves, crushed
3 tablespoons chopped fresh parsley
Salt and pepper to taste
About ¼ cup olive or vegetable oil
About 1 cup fine dry bread crumbs
About 1 cup grated Parmesan cheese

Rinse mushrooms off quickly or wipe with wet paper toweling to remove dirt. Pull off stems; cut off woody ends and finely chop mushrooms. Mix with ham, olives, almonds, garlic, and parsley. Season with salt and pepper. Brush mushroom caps with oil and fill with ham mixture. Sprinkle tops with bread crumbs, grated Parmesan cheese, and oil. Arrange in a shallow greased baking dish. Cook in preheated 375° F. oven for 15 minutes, or until tops are golden. Serves 12 as an appetizer or 6 as an entrée.

Anchovy-Stuffed Onions from Sicily

4 large yellow onions
Boiling salted water
12 anchovy fillets, minced
2 garlic cloves, crushed
⅓ cup chopped fresh parsley
1 teaspoon dried marjoram
¼ cup fine dry bread crumbs
6 tablespoons grated Parmesan cheese
Salt and pepper to taste
3 tablespoons olive or vegetable oil

Peel onions and blanch in boiling salted water for 15 minutes. Drain and rinse in cold water. Cut each onion in half crosswise and take out center, leaving a shell 4 rings thick. Chop finely the onion centers and combine in a bowl with anchovies, garlic, parsley, marjoram, bread crumbs, and 4 tablespoons grated cheese. Season with salt and pepper. Spoon mixture into onion halves. Sprinkle tops with remaining cheese and oil. Arrange in a well-buttered shallow baking dish. Bake, uncovered, in preheated 400° F. oven until onions are tender, about 30 minutes. Baste with juices once or twice during cooking. Serves 4 as a main dish or 8 as an accompaniment.

Polish Barley-Stuffed Cabbage Leaves

One of the world's most important foods, barley, has long been a vital ingredient in many staple European soups and other dishes. The grain has an attractive nutty flavor, is high in nutritive value, and makes a good stuffing for vegetables.

1 cup pearl barley	Salt and pepper to taste
1 large head green cabbage	1/3 cup chopped fresh parsley
Boiling salted water	1 1/2 cups beef bouillon
1 large onion, peeled and minced	2 tablespoons all-purpose flour
3 tablespoons vegetable oil	1 cup sour cream at room temperature
1/2 cup diced mushrooms	2 tablespoons chopped fresh dill
1 tablespoon fresh lemon juice	
1 teaspoon paprika	

Cook barley in water to cover for 1 hour, or until tender. Pull off and discard any wilted cabbage leaves. Cut out core to a depth of about 3 inches. Parboil cabbage in a large kettle of boiling salted water for about 20 minutes, or until leaves are soft enough to be pliable. Remove cabbage and carefully cut out tough part of rib from each leaf.

Sauté onion in heated oil until soft. Add mushrooms and lemon juice; sauté 3 minutes. Remove from heat and season with paprika, salt, and pepper. Add cooked barley and parsley, and mix well. Put one or more large spoonfuls of stuffing onto each leaf, the exact amount depending on the leaf's size. Roll up and fold in edges to securely enclose stuffing. Secure with toothpicks or tie with string. Place in a kettle. Cover with bouillon and cook slowly, covered, about 1 1/2 hours, or until leaves are tender. Remove stuffed leaves and keep warm. Combine 1 cup hot liquid with flour and cook slowly in a small saucepan, stirring, until thick and smooth. Add sour cream and dill and heat through. Pour sauce over stuffed leaves. Serves 8 to 10.

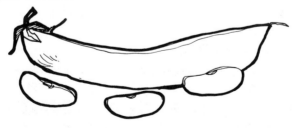

Mediterranean Stuffed Zucchini

4 small zucchini, washed	1 cup cooked rice
Boiling salted water	¼ cup chopped fresh parsley
1 large onion, peeled and minced	1 teaspoon dried oregano
2 garlic cloves, crushed	Salt and pepper to taste
7 tablespoons olive or vegetable oil	About ½ cup fine dry bread crumbs
2 medium-sized tomatoes, peeled and chopped	About ½ cup grated Parmesan cheese

Remove stem ends from zucchini. Put in a kettle of boiling salted water and cook 10 minutes. Remove with tongs and drain. Cut each zucchini in half lengthwise. Scoop out pulp and put it aside for later use in some other dish. Sauté onion and garlic in 3 tablespoons heated oil. Add tomatoes, rice, parsley, and oregano. Season with salt and pepper. Cook, stirring constantly, for 2 minutes. Remove from heat and spoon into drained zucchini shells. Sprinkle tops with bread crumbs, grated cheese, and remaining oil. Bake in preheated 350° F. oven for 15 minutes, or until tops are golden and crusty. Serves 8.

Greek Stuffed Grape Leaves with Lemon Sauce

Although the grape leaf is not actually a vegetable, it is a favorite for filling with savory stuffing. Grape leaves are often served as vegetables. Fresh tender leaves are best, but the most widely available types are those packed in brine and sold in jars or cans. These leaves, however, must be thoroughly drained and rinsed before they can be stuffed.

1 jar (16 fluid ounces) grape or vine leaves	1 cup uncooked rice
2 cups minced onions	1 cup chopped fresh parsley
1 pound ground meat (beef or lamb)	Salt and pepper to taste
	Boiling water
	¼ cup olive or vegetable oil

The Delectable Vegetable

Drain leaves and separate. Rinse each in warm water to remove any brine. Drain on paper toweling and set aside. Combine onions, meat, rice, and parsley. Season with salt and pepper. Work with hands to thoroughly mix ingredients. Shape into small balls, the exact size depending on size of leaf. Place 1 leaf on a flat surface, ribbed side up. Cut off any stem. Place a ball of stuffing near stem and fold sides of leaf over it. Roll up. Continue to fill leaves until all stuffing is used.

Place any torn and leftover leaves in a large heavy saucepan. Arrange stuffed leaves, one next to the other, in layers over them. Cover with boiling water and the oil. Cook slowly, tightly covered, about 40 minutes, or until stuffing is tender. Remove from heat and take stuffed leaves out with a slotted spoon. Reserve liquid and keep grape leaves warm. Serve with Lemon Sauce (below). Makes about 45. Serves 10 to 12 as an entrée.

Lemon Sauce

Beat 3 eggs until thick and light. Gradually add juice of 2 lemons and mix well. Beating constantly, add a little of hot reserved liquid in which stuffed grape leaves were cooked to egg-lemon mixture. Stir this into remaining hot broth and cook slowly, stirring, until thick and smooth. Remove from heat and serve at once; do not reheat.

Corn-Stuffed Pimientos Mexicana

3 tablespoons minced onion
¼ cup minced green peppers
2 tablespoons vegetable oil
1½ cups canned corn, drained, or frozen corn, defrosted
½ teaspoon celery seed
1 tablespoon chopped fresh parsley

Salt and pepper to taste
4 canned pimientos, drained
About ½ cup fine dry bread crumbs
About ½ cup grated Parmesan cheese
4 teaspoons butter or margarine

Sauté onion and peppers in oil in a skillet until tender. Mix with corn, celery seed, parsley, salt and pepper. Spoon into pimientos. Sprinkle tops with bread crumbs and cheese. Dot with butter. Arrange in a shallow baking dish and cook in preheated 400° F. oven for 15 minutes. Serves 4.

Italian Stuffed Artichokes

2 cups coarse dry bread crumbs	4 large globe artichokes
2 tablespoons minced anchovies	4 tablespoons olive or vegetable oil
½ cup chopped fresh parsley	1 lemon, sliced
½ cup grated Parmesan cheese	About 1 cup melted butter or margarine
3 to 4 garlic cloves, crushed	
Salt and freshly ground pepper to taste	

Combine first six ingredients and mix well. Thoroughly wash artichokes in cold water and drain. With a sharp knife cut off about ½ inch from each top. Cut and trim stem ends so artichokes will sit upright. With kitchen scissors snip off sharp leaf tips from each artichoke. Tear off and discard any loose or spotted leaves. With a teaspoon or knife, scoop out each choke. Spread out leaves gently with the hands. Spoon stuffing into center and between leaves of each artichoke. Fill a large kettle with 1 inch of water and bring to a boil. Place artichokes, one next to another, in it. Add oil and sliced lemon. Turn down heat and cook slowly, covered, about 45 minutes, or until artichokes are tender. Carefully remove with a slotted spoon and drain. Serve with melted butter. Serves 4.

Spanish Stuffed Tomatoes

Actually a fruit, the tomato is one of our most popular and versatile foods and is traditionally prepared and served as a vegetable. Tomatoes are a good source of nutrient minerals as they contain calcium and iron. They are also rich in vitamin C. Both raw and cooked tomato shells are ideal for stuffing with a diverse selection of savory mixtures.

8 medium-sized tomatoes	2 anchovy fillets, minced
Salt	5 tablespoons chopped fresh parsley
3 tablespoons olive or vegetable oil	¼ teaspoon dried thyme
2 tablespoons minced onion	Freshly ground pepper to taste
1 garlic clove, crushed	
2 cups minced cooked ham	

Cut a slice from stem end of each tomato. Scoop out pulp and mince it, discarding any tough pieces. Set aside. Lightly salt insides of tomato shells and turn over to drain. Heat 1 tablespoon oil in a skillet and sauté onion and garlic in it until tender. Add minced pulp and sauté 1 minute. Remove from heat and stir in ham, anchovies, parsley, and thyme. Season with salt and pepper. Spoon into drained tomato shells, packing firmly and rounding each into a mound at the top. Sprinkle remaining 2 tablespoons oil in a shallow baking dish and place stuffed tomatoes, standing upright in it. Cook, covered, in a preheated 350° F. oven about 25 minutes, or until tomatoes are just tender. Serves 8 as an appetizer or 4 as an entrée.

South American Eggplant-Filled Green Peppers

A very colorful and unusual stuffed vegetable dish.

About $1/2$ cup olive or vegetable oil	$1/4$ teaspoon dried oregano
	$1^1/2$ cups tomato sauce
1 medium-sized eggplant, about $1^1/4$ pounds, peeled and cut into $1/2$-inch cubes	Salt and pepper to taste
	4 medium-sized green peppers, seeded and cut in halves
1 garlic clove, crushed	$1/4$ cup grated Parmesan cheese

Heat 2 to 3 tablespoons oil in a medium skillet and sauté eggplant cubes, several at a time, until tender. Add more oil as needed. Add garlic, oregano, tomato sauce, salt and pepper. Mix well and cook, uncovered, 5 minutes. Parboil peppers in boiling salted water for 5 minutes. Drain. Fill with eggplant mixture. Sprinkle tops with cheese and bake, uncovered, in preheated 350° F. oven for 20 minutes. Serves 4.

Stuffed Mushrooms Provençale

- 1 pound fresh mushrooms, 2 to 3 inches in diameter
- About 5 tablespoons butter or margarine
- 1 tablespoon olive or vegetable oil
- ¼ cup minced shallots or green onions
- 2 garlic cloves
- 2 large tomatoes, peeled and chopped
- ⅓ cup chopped fresh parsley
- Salt and pepper, to taste

Rinse mushrooms quickly or wipe with wet paper toweling to remove any dirt. Pull off stems and cut off any tough woody ends. Mince stems and set aside. Brush caps with melted butter and arrange, hollow side up, in a shallow baking dish. Heat 2 tablespoons butter and the oil in a skillet and sauté shallots and garlic in it until tender. Add minced stems and tomatoes, and cook, stirring, 3 minutes. Remove from heat and stir in parsley. Season with salt and pepper. Spoon into mushroom caps, filling as full as possible. Bake in preheated moderate oven about 12 minutes, until mushroom caps are tender. Makes about 20, the number depending on size of mushrooms.

Cheese-Stuffed Cucumbers

- 3 medium-sized cucumbers
- 3 tablespoons butter or margarine
- ½ cup chopped onion
- 2½ cups soft bread cubes
- ½ cup grated Parmesan cheese
- ⅓ cup chopped fresh parsley
- Salt and pepper to taste
- ½ cup Mozzarella cheese cubes

Peel cucumbers and cut in halves lengthwise. Scoop out and discard seeds and pulp. Melt butter in a small skillet. Add onions and sauté until tender. Remove from stove and mix with bread cubes, Parmesan cheese, parsley, salt and pepper. Spoon into cucumber shells and top with cubes of Mozzarella. Put in a shallow baking dish and add ½ cup water. Bake in preheated 375° F. oven about 45 minutes, until cucumbers are tender. Serves 6.

Greek Rice-Stuffed Tomatoes

8 medium-sized ripe tomatoes of uniform size
Salt
1 cup minced onion
6 tablespoons olive or vegetable oil
¼ cup chopped pine nuts
1 cup raw rice
2 cups chicken bouillon
¼ cup currants or chopped raisins
1 teaspoon ground cinnamon
Pepper to taste
3 tablespoons chopped fresh dill

Cut a slice from top of each tomato. Scoop out pulp and seeds and reserve. Sprinkle insides of tomatoes with salt and invert to drain. Sauté onion in heated oil in a skillet until tender. Add pine nuts and rice, and cook over fairly high heat for 5 minutes, or until rice is translucent. Add reserved tomato pulp, chicken bouillon, currants, and cinnamon. Season with salt and pepper. Cook slowly, covered, about 20 minutes, or until rice grains are tender and liquid is absorbed. Stir in dill. Remove from heat and cool. Spoon mixture into tomato shells, filling lightly. Arrange in a greased baking dish. Add ½ inch water and bake in preheated 350° F. oven 30 minutes, or until tender. Serve hot or cold. Serves 8.

South Seas Stuffed Sweet Potatoes

4 large sweet potatoes
2 tablespoons butter or margarine
3 tablespoons pineapple juice
2 tablespoons brown sugar
½ teaspoon ground nutmeg
⅓ cup chopped blanched almonds

Scrub potatoes well and dry thoroughly. Bake in preheated 350° F. oven until tender when pierced with a fork, 40 to 50 minutes. Remove from oven and cut a slice from top of each. Scoop out insides while still hot, being careful not to break skins, and mash well. Combine with remaining ingredients and spoon into potato shells. Bake in 450° F. oven 10 minutes. Serves 4.

Stuffed Potatoes from Idaho

One of the world's greatest white potatoes is the large one grown in the volcanic soil of southern Idaho and named for that state. It is highly prized for baking and excellent for stuffing.

4 medium-sized Idaho potatoes	3 tablespoons butter or margarine
3 tablespoons minced onion	1/4 cup grated Parmesan cheese
3 tablespoons minced green pepper	Salt and pepper to taste
3 tablespoons milk	

Scrub potatoes well and dry thoroughly. Bake in a preheated 425° F. oven until very tender when pierced with a fork, 50 to 60 minutes. Scoop out insides while still hot, being careful not to break skins, and mash well. Mix with onion, green pepper, milk, 2 tablespoons butter, cheese, salt and pepper. Spoon into shells and dot tops with remaining tablespoon of butter. Return to oven and bake 15 minutes longer. Serve at once. Serves 4.

Oregon Salmon-Stuffed Tomatoes

6 medium-sized ripe tomatoes	1 cup chopped canned salmon
Salt	1 tablespoon lemon juice
1/4 cup minced onions	1/2 teaspoon dried oregano
1 garlic clove, crushed	Freshly ground pepper to taste
1 tablespoon olive or vegetable oil	About 1/2 cup fine dry bread crumbs
6 anchovy fillets, minced	
3 tablespoons drained capers	

Cut a slice from top of each tomato. Carefully scoop out pulp and seeds. Chop pulp and reserve. Sprinkle insides of tomatoes lightly with salt and invert to drain. Sauté onions and garlic in heated oil until tender. Add reserved tomato pulp, anchovies, capers, salmon, lemon juice, and oregano. Season with pepper. Mix well and cook 1 minute. Remove from heat and spoon into tomato shells. Sprinkle with bread crumbs. Arrange in a greased shallow baking dish and bake in preheated 375° F. oven until tomatoes are cooked, about 20 minutes. Serves 6.

Persian Rice-Filled Grape Leaves

1 jar (16 fluid ounces) grape or vine leaves	1 teaspoon sugar
$1/3$ cup olive or vegetable oil	Salt and pepper to taste
1 cup minced onion	3 tablespoons chopped fresh parsley
1 cup raw rice	Juice of 1 lemon
$1/4$ cup pine nuts	2 tablespoons tomato paste
1 cup hot beef bouillon or water	Lemon wedges

Drain leaves and separate. Rinse each in warm water; drain on paper toweling and set aside. Heat oil in a large skillet and sauté onion in it until tender. Add rice and pine nuts, and sauté, stirring constantly, until rice becomes translucent, about 5 minutes. Add hot bouillon and sugar. Season with salt and pepper. Stir well. Cook slowly, covered, about 15 minutes, or until rice grains are tender and liquid is absorbed. Remove from heat and stir in parsley. Cool.

Place a leaf, ribbed side up, on a flat surface and cut off stem, if any. Place a spoonful of stuffing near stem end and fold sides of leaf over it. Roll up. The amount of stuffing will depend on size of leaf. Continue to fill leaves until all stuffing is used. Place any leftover vine leaves in a large saucepan. Arrange stuffed leaves, one next to the other, in layers over them. Heat 2 cups water in skillet in which stuffing was cooked, stirring in any drippings. Add lemon juice and tomato paste and stir well. Bring to a boil and pour over stuffed leaves. Cook slowly, tightly covered, about 40 minutes, or until stuffing is fully cooked. Add more water while cooking, if necessary. Carefully remove from pan with a slotted spoon. Serve cold with lemon wedges. Makes about 45.

Main Dishes

In the United States we call the part of the meal that contains the most substantial nourishment the main dish. At dinner this is customarily a dish made of meat, poultry, or seafood, but certainly one starring vegetables, with or without the aforementioned foods, can be an excellent choice.

Very often the main dish is called the entrée. This term is particularly used in restaurants. Yet the word literally means "entrance," and in England and France it is used to indicate a third course served after the soup and the fish and before the meat.

Regardless of what it is called, the main dish of a luncheon, supper, or dinner is very important; the star of the meal, it should be carefully prepared and served.

The main dishes in this section have been chosen to offer the cook something different in the way of selection. While not unique, they are not commonplace; whether prepared with vegetables alone or combined with meat, poultry, or seafood, they represent worldwide culinary contributions.

North African Vegetable Couscous

Staple stews in North Africa called *couscous* are made with a grain of the same name, a medley of vegetables, and very often meat or poultry. Traditionally the stew is cooked in a *couscoussière*, a type of large double boiler that has a top with a perforated bottom placed over a round pan. The grain is sold in packages in the United States at some supermarkets and most specialty food stores. Cubes of lamb or pieces of chicken may be added to this recipe, if desired. Brown in oil and cook partially before adding the vegetables.

- 1 package (500 grams or 17 ounces) couscous
- ½ cup water
- 2 large onions, peeled and sliced
- ¼ cup olive or vegetable oil
- 1 tablespoon crushed red pepper
- ½ teaspoon ground cumin
- Salt and pepper to taste
- 1 large can (1 pound, 12 ounces) tomatoes, undrained
- 3 medium-sized zucchini, washed and sliced
- 2 small white turnips, peeled and cubed
- 2 large green peppers, cleaned and cut into strips
- 3 medium-sized carrots, scraped and cut into 1-inch pieces
- 1 can (1 pound, 4 ounces) chickpeas, drained
- ¼ cup chopped fresh coriander or parsley

Spread *couscous* over a tray and sprinkle with water. Mix with the fingers. When ready to cook, put in a colander lined with cheesecloth or in top of a *couscous* steamer. Sauté onions in heated oil in a large kettle or in bottom of a *couscous* steamer until soft. Add red pepper, cumin, salt and pepper, and cook 1 minute. Stir in tomatoes and cook another minute. Add vegetables and a little water, about 1 cup. Put colander or steamer top with the *couscous* over the vegetables. Cook, covered, about 30 minutes, or until ingredients are tender. Add coriander. To serve, spoon *couscous* onto a large platter and arrange vegetables and liquid over and around it. Serves 6.

Note: Check during cooking to see if more water should be added to vegetables. There must be a fair amount of liquid because the *couscous* absorbs considerable liquid from the vegetables.

Vegetable Khoreshe from Iran

Khoreshe, a thick well-flavored "sauce," similar to a stew, is traditionally prepared in Iran with fresh or dried vegetables, meat or poultry, or fresh or dried fruit; it is served over rice as a main course. This is one of several kinds.

3	tablespoons butter or margarine	1	cup chopped celery
1	large onion, peeled and chopped	1	cup chopped fresh parsley
		2	tablespoons fresh lemon juice
2	leeks, white parts only, washed and sliced	1/2	teaspoon ground cinnamon
		1/4	teaspoon ground nutmeg
1	cup chopped scallions, with tops		Salt and pepper to taste
			About 1 cup water
2	cups cut-up spinach leaves	1	cup plain yogurt at room temperature
2	cups cut-up green beans		
1	medium-sized eggplant, washed and cubed		

Heat butter or margarine in a large saucepan or kettle. Add chopped onion, leeks, and scallions. Sauté until tender. Add remaining ingredients, except yogurt. Cook slowly, covered, about 25 minutes, or until vegetables are tender. Add more water while cooking, if needed. Mix in yogurt and leave on the stove long enough to heat through. Spoon over hot cooked rice at the table. Serves 4 to 6.

Korean Chop Chay

A favorite and excellent main dish in Korea is named *chop chay* and is a combination of chopped foods that are cooked very quickly and well seasoned with the traditional soy sauce and sesame seeds. It is often made of vegetables only, but it can also include strips of lean beef.

	About ¼ cup peanut or vegetable oil	4	stalks celery, cleaned and sliced
4	medium-sized carrots, scraped and finely chopped	4	bamboo shoots, chopped
10	scallions, cleaned and cut into 1-inch lengths	2	cans (1 pound each) bean sprouts, drained
2	large onions, peeled and finely chopped	2	garlic cloves, crushed
		⅓	cup soy sauce
2	cups washed, trimmed, and chopped spinach leaves	2	tablespoons toasted sesame seeds * (optional)
½	pound fresh mushrooms, cleaned and chopped	½	pound cellophane noodles or vermicelli, cooked and drained

Heat 2 tablespoons oil in a skillet and sauté carrots and scallions and onions in it for about 2 minutes; they should be a little crisp. Remove to a plate and keep warm. Add more oil to skillet and sauté spinach, mushrooms, celery, bamboo shoots, and bean sprouts in it for about 4 minutes. Return carrots and onions to skillet and add garlic, soy sauce, and sesame seeds. Cook slowly a few minutes to blend flavors. Mix in noodles and leave on low heat, stirring, until mixture is hot. Serves 4 to 6.

* To toast sesame seeds, place in a pan over low heat and cook until nut-brown.

Eggplant alla Parmigiana

This well-known Italian eggplant specialty derives its name from Parmesan cheese, *parmigiano*. It is a good buffet dish.

2	medium-sized eggplants, about 1¼ pounds each.	1	teaspoon dried basil
		½	teaspoon dried oregano
	Salt		Pepper to taste
	About 1 cup olive or vegetable oil		About ¾ cup all-purpose flour
		2	to 3 eggs, beaten
1	medium-sized onion, peeled and minced	¾	pound Mozzarella cheese, sliced
1	can (4 ounces) tomato paste		About 1 cup grated Parmesan cheese
2¼	cups water		

Wash eggplants; cut off stems; cut crosswise into slices about ¼ inch thick. Put in a colander and sprinkle with salt. Leave to drain for 30 minutes. Meanwhile, heat 2 tablespoons oil in a skillet and add onion. Sauté until tender. Mix in tomato paste, water, and herbs, and season with salt and pepper. Cook slowly over low heat, uncovered, about 20 minutes, stirring occasionally. Drain eggplant slices and dry thoroughly. Dust each with flour and dip in beaten egg. Fry in hot oil on both sides until golden. Add more oil as needed. Drain on absorbent paper. Line a shallow baking dish with a little of the tomato sauce. Arrange a layer of eggplant slices over it. Cover with a layer of Mozzarella slices, more sauce, and a sprinkling of Parmesan cheese. Repeat layers until all ingredients are used. Top with Parmesan cheese. Bake in preheated 350° F. oven for 30 minutes. Serves 6.

Fish-Vegetable Plaki from Greece

Whole or cut-up fish that is baked on a bed of vegetables is called *plaki* in Greece. It may be served hot or cold and is a great favorite at the colorful Greek *tavernas*—informal restaurants—noted for their gaiety and good food.

1 4-pound whole fish, dressed; or 3 pounds fish fillets (bass, pike, halibut, flounder)	3 large tomatoes, peeled and sliced
Salt and pepper to taste	3 carrots, scraped and thinly sliced
2 large onions, peeled and sliced	1 cup sliced celery
3 medium-sized leeks, white parts only, washed and sliced	2 bay leaves
2 garlic cloves, crushed	1 teaspoon dried oregano
⅓ to ½ cup olive or vegetable oil	2 medium-sized lemons, sliced
	⅓ cup chopped fresh parsley

Wash fish and wipe dry. Sprinkle inside and out with salt and pepper. Sauté onions, leeks, and garlic in heated oil in a skillet until tender. Add tomatoes, carrots, celery, bay leaves, oregano, and season with salt and pepper. Sauté 10 minutes. Spoon into a shallow baking dish and spread evenly. Arrange fish over vegetable mixture. Place lemon slices over fish. Bake in a preheated 350° F.

oven about 30 minutes, or until fish flakes with a fork. Sprinkle with parsley 10 minutes before cooking is finished. The baking time will be longer for whole fish than for fillets. Serves 4 to 6.

Baked Corn-Bean–Filled Enchiladas

This Mexican dish made with stuffed rolled tortillas is an equally good choice for brunch, luncheon, or supper.

2 cans (8 ounces each) tomato sauce	2 cups drained, cooked or canned corn
2 garlic cloves, crushed	2 cups drained, cooked or canned pinto or red beans
1 teaspoon dried oregano	
2 to 4 teaspoons chili powder	2 medium-sized onions, peeled and chopped
Salt and pepper to taste	
16 prepared or canned tortillas	1½ cups grated Monterey Jack or Cheddar cheese
About ⅔ cup vegetable oil	

Combine tomato sauce, garlic, oregano, chili powder, salt and pepper, in a small saucepan. Bring to a boil. Reduce heat and cook slowly, uncovered, 10 minutes. Fry tortillas, one at a time, on both sides in heated oil in a skillet until limp. Spoon one heaping tablespoon of corn and one of beans along center of each tortilla. Sprinkle with chopped onion and roll up around filling. Arrange, seam sides down, next to each other in a shallow greased baking dish. Spoon sauce over enchiladas and sprinkle with cheese. Bake in a preheated 350° F. oven until enchiladas are hot and cheese has melted, about 25 minutes. Serves 4.

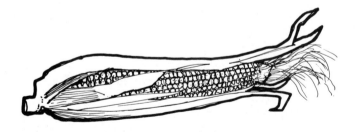

Japanese Vegetable Sukiyaki

Although most versions of this popular Oriental dish are made with thin slices of beef, as well as vegetables, this one includes only the latter. Cook in a special sukiyaki pan, electric skillet, or ordinary large skillet. Sukiyaki is traditionally prepared in front of diners at the table. Thus ingredients are prepared beforehand, arranged attractively on a platter, and brought to the table when ready to cook.

- 1/2 cup soy sauce
- 1/4 cup sake or dry white wine
- 1/4 cup dashi (Japanese soup stock) or water
- 1/2 teaspoon monosodium glutamate
- 2 large onions, peeled, cut in halves and sliced thinly
- 8 scallions, cleaned and cut into 2-inch pieces
- 1 cup sliced bamboo shoots
- 2 cups Chinese cabbage, cut in diagonal slices
- 2 cups fresh spinach, washed, trimmed, and chopped
- 1 cup fresh or frozen defrosted snow peas
- About 1/3 cup peanut or vegetable oil
- 10 1-inch cubes bean curd

Combine first four ingredients in a pitcher and set aside. Prepare vegetables and arrange on a platter. When ready to cook, heat 2 tablespoons oil in cooking utensil. With chopsticks or a fork, add vegetables, about one fourth of them at a time, and sauté, turning, until cooked just a bit. Add bean curd and then soy-sauce mixture, and cook, stirring, until vegetables are tender. Do not overcook as they should be a little crisp. Serve at once. Serves 4.

Romanian Ghiveciu

A most attractive and delectable Romanian dish is made with a beautiful variety of baked vegetables, sometimes as many as twelve or more. There is no one particular recipe for *ghiveciu* as every cook prepares it according to instinct or family tradition. Cauliflower, celery root, cabbage, green peas, leeks, squash, and/or turnips can be added to the following ingredients, if desired.

2 medium-sized onions, peeled and sliced
About 1½ cups olive or vegetable oil
2 medium-sized potatoes, peeled and sliced
1 medium-sized unpeeled eggplant, about 1¼ pounds
1 large green pepper, cleaned and cut into cubes
2 large carrots, scraped and sliced
1 cup fresh or frozen cut-up green beans
1 cup hot vegetable bouillon or water
2 garlic cloves, crushed
¾ cup chopped fresh parsley
¼ cup mixed fresh herbs
Salt and pepper, to taste
4 medium-sized tomatoes, peeled and sliced
¼ pound seedless green grapes, washed

Sauté onion slices in ⅓ cup heated oil until tender. Take out and set aside. Add potato slices; sauté until just tender. Take out and set aside. Repeat for all vegetables, except green beans. Arrange sautéed vegetables and green beans in an extra large, round or rectangular, shallow baking dish. Combine bouillon, garlic, parsley, and herbs; pour over ingredients. Add any drippings from pan. Season with salt and pepper. Bake, covered, in a preheated 350° F. oven until vegetables are cooked, about 30 minutes. Arrange tomatoes and grapes over top 15 minutes before end of cooking. The dish can be eaten hot or cold. Serves 6.

Tomato-Zucchini Lasagna

Cooked broad, flat noodles, baked in layers with zucchini and topped with a flavorful tomato sauce, make an interesting flavorful dish for a buffet or dinner.

2 quarts water
1 teaspoon olive or vegetable oil
1 teaspoon salt
½ pound lasagna macaroni
4 zucchini, washed and stems removed
1 can (6 ounces) tomato paste
1 cup tomato juice
½ cup hot water
1 garlic clove, crushed
1 teaspoon dried oregano
½ teaspoon dried basil
Salt and pepper to taste
2 cups chopped onions
About ½ cup grated Parmesan cheese

Heat water, oil, and 1 teaspoon salt to boiling in a large kettle. Add lasagna and cook for 12 to 15 minutes; drain well. Arrange on a platter. Meanwhile, cut unpeeled zucchini into ¼-inch slices and fry in heated oil in a skillet. Remove to a plate. Combine remaining ingredients, except cheese, in a saucepan and cook slowly, uncovered, for 10 minutes. Arrange a layer of lasagna in a shallow rectangular greased baking dish. Cover with a layer of zucchini slices and several spoonfuls of sauce. Sprinkle with Parmesan cheese. Repeat until all ingredients are used, ending with cheese on top. Bake in preheated 350° F. oven for 30 minutes, or until bubbly and cooked. Serves 8.

Latin American Vegetable Stew

This flavorful one-dish meal, called *locro,* is sometimes made with cubes of beef, in addition to its medley of vegetables.

- 3 tablespoons vegetable oil
- 1 large onion, peeled and chopped
- 1 garlic clove, crushed
- 1 medium-sized green pepper, cleaned and chopped
- 4 medium-sized carrots, scraped and sliced
- 4 medium-sized white potatoes, peeled and cubed
- 2 sweet potatoes, peeled and cubed
- 2 cups cubed squash or pumpkin
- 2 cups fresh, frozen, or canned whole corn kernels
- 1 cup fresh or frozen green peas
- 1 teaspoon ground hot red pepper
- Salt and pepper to taste
- 2 cups water
- ¼ cup chopped fresh parsley

Heat oil in a large kettle; sauté onion, garlic, and green pepper in it until tender. Add remaining ingredients, except parsley. Bring to a boil. Reduce heat and simmer slowly about 40 minutes, or until vegetables are tender. Check during cooking to see if more water is needed. Stir in parsley. Serves 4 to 6.

Swedish Vegetable-Rice Pot

3 slices bacon, chopped	2 cups chopped green cabbage
1 large onion, peeled and chopped	¾ cup long-grain rice
	1½ cups tomato juice
3 medium-sized stalks celery, cleaned and chopped	½ teaspoon dried thyme
	1 bay leaf
3 medium-sized carrots, scraped and thickly sliced	Salt and pepper to taste
	1 cup frozen green peas
1 medium-sized rutabaga or white turnip, peeled and cubed	½ cup sour cream
	2 tablespoons chopped fresh dill

Fry bacon in an extra large saucepan to release fat. Add onion, celery, and carrots, and sauté 5 minutes. Add rutabaga, cabbage, rice, tomato juice, thyme, bay leaf, salt and pepper. Cook covered over moderate heat for 20 minutes. Add peas and cook another 10 minutes. Stir in sour cream and dill. Cook over low heat 5 minutes. Serves 4 to 6.

New England Boiled Dinner

This is the most famous of the colonial one-dish meals that were prepared weekly for the family dinner table and made in the early days with root vegetables stored in the cellar. It is hearty and nourishing.

4 to 5 pounds lean corned brisket of beef	6 medium-sized potatoes, peeled
	2 pounds parsnips, peeled and cubed
6 medium-sized carrots, scraped and halved	Salt and pepper to taste
1 medium-sized green cabbage, cut into wedges with core removed	6 medium-sized beets, washed, with tops cut off about 2 inches from beets
2 pounds yellow turnip or rutabaga, peeled and cut in thick slices	Boiling salted water

Put corned beef in a large kettle with water to cover. Simmer, covered, for 3 hours, skimming from time to time. Add carrots, cabbage, turnip, potatoes, and parsnips. Season with salt and pepper. Continue cooking slowly, covered, until vegetables are tender, about thirty minutes. Meanwhile, cook beets separately in boiling salted water until tender, about 35 minutes, depending on their size. When cooked, drain in colander, slip off skins, and keep warm. To serve the dinner, slice the corned beef and arrange the cooked vegetables around it. Serves 6 to 8.

Red Flannel Hash

This favorite early American supper dish was made with leftovers from the New England Boiled Dinner. Sometimes the corned beef was omitted and only vegetables were used. The cabbage was fried separately and served with the hash.

2 cups chopped cooked corned beef	1 medium-sized onion, peeled and chopped
2 cups chopped boiled peeled potatoes	1/4 cup light cream or milk
	Salt and pepper to taste
1 cup chopped cooked skinned beets	3 tablespoons bacon drippings or butter

Combine beef, vegetables, cream, salt and pepper, in a bowl. Heat drippings or butter in a skillet. Add meat-vegetable mixture and heat slowly, occasionally shaking the skillet and loosening ingredients around the edges. When a brown crust has formed on the bottom, turn onto a warm platter. Fold as for an omelet, if desired. Serves 4.

Broccoli Indienne in Patty Shells

This is a flavorful easy-to-prepare main dish for luncheon or dinner.

1 package (10 ounces) frozen broccoli spears	Salt and pepper to taste
1 tablespoon butter or margarine	²/₃ cup chicken broth
2 tablespoons all-purpose flour	½ cup light cream or milk
1 teaspoon turmeric powder	1 cup canned or cooked tiny white onions
¼ teaspoon paprika	4 hot pastry shells

Cook broccoli according to package directions until just tender. Drain and cut into bite-size pieces. Melt butter in a saucepan; stir in flour and cook 1 minute. Add turmeric powder, paprika, salt and pepper, and cook another minute. Gradually add broth, stirring as adding, and cook until thick and smooth. Add cream and continue to cook slowly. Mix in onions and broccoli, and leave on stove until heated. Just before serving, spoon into hot pastry shells. Serves 4.

Ratatouille of Provence

This savory Mediterranean dish, prepared with a medley of vegetables and seasonings, can be served as an appetizer or for the main course, hot or cold. It is an attractive buffet dish.

1 medium-sized eggplant, about 1¼ pounds, washed	
Salt	1 large sweet green pepper, cleaned and cut into strips
About 1 cup olive or vegetable oil	3 large tomatoes, peeled and chopped
2 medium-sized onions, peeled and chopped	Freshly ground black pepper
2 zucchini, about ½ pound each, washed, stemmed, and sliced	¼ teaspoon dried basil
	¼ teaspoon dried thyme or marjoram
2 or 3 garlic cloves, crushed	Juice of 1 lemon

Peel eggplant and cut into slices about ¼ inch thick. Put in a colander and sprinkle with salt. Leave to drain 30 minutes. Wipe dry. Heat ⅓ cup oil in a large skillet. Add onions and simmer until soft. Push aside and add eggplant slices, several at a time, and cook slowly until tender, adding more oil as needed. Remove to a plate. Add zucchini and cook until tender. Return eggplant to pan. Add remaining ingredients, except lemon juice. Simmer, covered, about 30 minutes, until ingredients are tender. Add lemon juice and cook, uncovered, 5 minutes. Serves 4 to 6.

Dinner Vegetable Loaf

¼ cup butter or margarine	1 cup cooked pinto or kidney beans
1 medium-sized onion, peeled and chopped	2 eggs, beaten
¼ cup chopped green pepper	½ teaspoon dried oregano or thyme
¼ cup chopped celery	⅓ cup chopped fresh parsley
2 cups soft bread cubes	Salt and pepper to taste
1 cup cooked corn	
1 cup cooked green peas	

Melt butter in a medium skillet. Add onion, green pepper, and celery; sauté 2 to 3 minutes; turn into a large bowl. Add remaining ingredients and mix to combine well. Spoon into a greased loaf pan, 9 by 5 by 3 inches, and bake in a preheated 350° F. oven until done, about 35 minutes. To serve, cut into slices. Serve with a tomato or mushroom sauce, if desired. Serves 4 to 6.

Hungarian Vegetable Goulash

A meatless version of the famous Hungarian goulash that can be served over wide cooked noodles for luncheon or supper or an evening get-together.

2 tablespoons shortening or vegetable oil	4 medium-sized potatoes, peeled and cubed
2 large onions, peeled and chopped	2 cups cut-up green beans
	Salt and pepper to taste
2 to 3 tablespoons paprika	About 1 cup water
1 large green pepper, cleaned and chopped	2 tablespoons chopped fresh dill
2 medium-sized carrots, scraped and sliced	1 cup sour cream at room temperature
3 medium-sized tomatoes, peeled and chopped	

Heat shortening or oil in a medium saucepan and sauté onions in it until soft. Add paprika and cook 1 minute. Mix in green peppers and carrots, and sauté 2 minutes. Add remaining vegetables, salt and pepper, and water, and cook slowly, covered, about 30 minutes, or until vegetables are tender. Add a little more water during cooking, if necessary. Mix in dill and sour cream. Serves 4.

Ceylonese Vegetable Curry

Serve for luncheon or supper, or double the recipe and serve at a buffet.

2 to 3 tablespoons peanut or vegetable oil	1 cup cut-up green beans
	1 cup chopped scraped carrots
2 large onions, peeled and chopped	1 cup green peas
	1 cup cubed peeled potatoes
2 leeks, white parts only, washed and sliced	3 tomatoes, peeled and cut in wedges
1 or 2 garlic cloves, crushed	1 cup boiling water
1 tablespoon curry powder	2 tablespoons grated fresh or frozen coconut
1 to 2 teaspoons chili powder	
1 teaspoon ground cumin	2 tablespoons fresh lime or lemon juice
Salt and pepper to taste	

Heat oil in a skillet. Add onions, leeks, and garlic, and sauté until tender. Push aside and add curry, chili powder, cumin, salt and pepper. Cook 1 minute. Add remaining ingredients, except lime or lemon juice, and cook slowly, covered, about 25 minutes, until vegetables are cooked but not mushy. Add lime or lemon juice. Serve over hot boiled rice with chutney or relish, if desired. Serves 4.

Polish Mushroom-Filled Pancakes

An excellent main dish for a luncheon or brunch.

Pancakes
- 1 cup milk
- 1 large egg
- 1 cup sifted all-purpose flour
- 1/4 teaspoon salt
- About 3 tablespoons butter or margarine
- Mushroom Filling (see below)
- About 1 cup sour cream at room temperature

Combine milk and egg in a bowl; mix with a whisk or fork. Stir in flour and salt; beat well. Heat butter or margarine in a lightly greased 7- or 8-inch skillet, and spoon 3 tablespoons batter into it, pouring all at once from a small glass. Tip pan immediately to spread batter evenly. Cook until underside is done and bubbles form on top. Turn over with a spatula, and cook on other side. Slip onto a warm plate and keep warm in a preheated 250° F. oven while cooking other pancakes. When all are cooked, put a large spoonful of Mushroom Filling lengthwise along each one. Roll up. Serve at once topped with a spoonful of sour cream. Serves 10 to 12.

Mushroom Filling
- 1 pound fresh mushrooms
- 6 scallions, minced
- 3 to 4 tablespoons butter or margarine
- 1/2 cup sour cream at room temperature
- 1 tablespoon all-purpose flour
- 2 tablespoons chopped fresh dill or parsley
- Salt and pepper to taste

Clean mushrooms by rinsing quickly or wiping with wet paper toweling to remove any dirt. Sauté scallions in heated butter until tender. Add mushroom slices and sauté 4 minutes. Stir in remaining ingredients and leave on stove long enough to heat through.

Chinese Sub Gum Chow Mein

A *sub gum* chow mein is one that includes a special or elaborate combination of ingredients—generally seafood, poultry, and vegetables—served over noodles.

½	pound Chinese or packaged fine egg noodles	½	cup sliced bamboo shoots
	Boiling salted water	1	cup drained bean sprouts
	About ⅓ cup peanut or vegetable oil	2	cups sliced fresh mushrooms
		1	package (6 ounces) frozen snow peas or green peas, defrosted
2	cups diced cooked chicken		
1	large onion, peeled and sliced thinly	2	tablespoons cornstarch
		2	tablespoons soy sauce
2	cups chopped, cleaned Chinese cabbage	2	teaspoons sugar
		1	cup chicken broth
2	cups chopped, cleaned spinach		

Cook noodles in boiling salted water and drain. Heat 2 tablespoons oil in a large skillet and fry noodles until crisp. Heat 3 tablespoons oil in a skillet and sauté chicken about 1 minute. Push aside. Add onion and sauté until tender. Stir in cabbage, spinach, bamboo shoots, bean sprouts, mushrooms, and peas. Stir-fry about 4 minutes, until just tender. Combine remaining ingredients and pour into skillet. Leave on stove long enough to heat through and blend flavors. To serve, spoon over warm noodles. Serves 6.

Austrian Egg-Noodle Ring with Creamed Fresh Asparagus

A good main-course dish for a women's luncheon; it can also be made with frozen asparagus as a substitute for the fresh vegetable, if desired.

12 ounces medium-wide egg noodles	Salt and pepper to taste
3 tablespoons butter or margarine, melted	2 cups sour cream at room temperature
3 eggs, separated	Creamed Fresh Asparagus (see below)
2 tablespoons chopped fresh parsley	

Cook noodles until soft; drain well. Turn into a large bowl. Combine melted butter and egg yolks, parsley, salt and pepper, and mix well. Add sour cream. Mix with noodles, tossing with two forks to combine well. Beat egg whites until stiff and fold into noodle mixture. Spoon into a greased 6½-cup ring mold. Place in a pan of hot water. Bake in preheated 375° F. oven until firm, about 40 minutes. Unmold on a warm platter. Fill center and adorn top of ring with warm creamed vegetables. Serves 8 to 10.

Creamed Fresh Asparagus

1 pound fresh asparagus	2 cups light sweet cream or sour cream
Salted water	
2 tablespoons butter or margarine	Dash of grated nutmeg
	Salt and pepper to taste
¼ cup all-purpose flour	

Wash and clean asparagus. Slice diagonally into 1½-inch pieces. Put pieces, except tips, with ½ inch salted water in a saucepan and bring to a boil. Lower heat and cook, covered, 2 minutes. Add tips and cook until tender, about 6 minutes. Remove from heat and drain. Melt butter in a saucepan. Mix in flour and cook 1 minute. Stir in cream and cook slowly, stirring, until thick and smooth. Add asparagus and season with nutmeg, salt and pepper.

Vegetable Stew from Istanbul

This superb Turkish favorite, called *türlü*—gardeners' stew—is generally made with lamb and a medley of whatever vegetables are in season. It is a traditional summer dish.

2 medium-sized onions, peeled and sliced	4 large tomatoes, peeled and chopped
2 large carrots, scraped and sliced	2 medium-sized green peppers, cleaned and chopped
1/3 cup olive or vegetable oil	4 medium-sized potatoes, peeled and cubed
4 medium-sized zucchini, washed, stems removed, and sliced	1/2 teaspoon dried thyme
	1/4 cup chopped fresh parsley
1 pound fresh green beans, stemmed and cut-up	Salt and pepper to taste
	About 1 cup water
1/2 pound fresh okra, stemmed and cut-up	

Sauté onions and carrots in heated oil in a large saucepan for 5 minutes. Add remaining ingredients and mix well. Cook slowly, covered, until vegetables are tender, about 35 minutes. Check during cooking to see if a little more water is needed. Serves 6 to 8.

Indonesian Bahmi Goreng

This flavorful vegetable-noodle dish is traditionally made with pork and/or shrimp, but the pork in this recipe may be left out, if desired.

1/2 pound fine egg noodles or vermicelli	About 8 tablespoons peanut or vegetable oil
Boiling salted water	1 cup chopped cleaned celery
1 1/2 pounds boneless pork strips, without fat	2 teaspoons minced ginger root
	1 cup chopped scraped carrots
1/2 cup soy sauce	1 cup drained bean sprouts
2 garlic cloves, crushed	3 cups chopped Chinese cabbage
2 large onions, peeled and sliced thinly	Freshly ground pepper
2 eggs, beaten	

Cook noodles in boiling salted water until tender. Drain. Spread out on a large plate and cool. Refrigerate 2 hours. Put pork strips, 1/3 cup soy sauce, garlic, and onions in a large bowl. Mix well. Leave to marinate for 2 hours. Pour eggs into a lightly greased skillet and tilt at once to spread evenly. Cook over low heat until set. Remove to a plate and cool. Cut into strips and set aside. When ready to cook, heat 2 tablespoons oil in a large skillet. Add pork mixture and sauté until cooked. Remove to a plate. Heat 2 more tablespoons oil in skillet. Add celery and ginger root, and sauté until soft. Add 1 tablespoon more oil and vegetables, and sauté until tender. Return cooked pork and onions to skillet. Stir in remaining soy sauce and season with pepper. Leave over low heat. Heat 3 tablespoons oil in another skillet. Add chilled noodles and cook until crisp and golden. Remove with a slotted spoon and mix with pork-vegetable mixture. Garnish with cooked egg strips. Serve 4 to 6.

Greek Eggplant Moussaka

This well-known Middle Eastern dish is made in various ways, but usually it is a savory meat mixture used as a filling between sliced eggplant or potatoes. It is a good buffet dish.

2 medium-sized onions, peeled and chopped	$1/2$ cup water
1 or 2 garlic cloves, crushed	2 medium-sized eggplants, about $1^{1}/_{4}$ pounds each, washed
About 1 cup olive or vegetable oil	
1 pound ground lean beef	$1/4$ cup butter or margarine
2 tablespoons tomato paste	$1/4$ cup all-purpose flour
$1/2$ teaspoon dried oregano	2 cups hot milk
$1/4$ cup chopped fresh parsley	2 eggs, lightly beaten
Salt and pepper to taste	$1/2$ cup grated Parmesan cheese
	Pinch of grated nutmeg

Sauté onions and garlic in 2 tablespoons heated oil in a skillet until tender. Add meat and cook, separating with a fork, until redness disappears. Add tomato paste, oregano, parsley, salt and pepper, and water. Cook, uncovered, for 20 minutes. Meanwhile, cut unpeeled eggplant into thin lengthwise slices. Fry eggplant in heated oil, a few slices at a time, until tender and golden on both sides. Put alternate layers of eggplant and meat mixture in a shallow 2-quart baking dish, beginning and ending with a layer of eggplant. Melt butter in a saucepan and stir in flour. Cook 1 minute. Gradually add milk, stirring constantly. Season with salt and pepper. Remove from heat. Add eggs, cheese, and nutmeg, and mix well. Carefully turn into the baking dish to cover the top layer of eggplant. Bake in preheated 350° F. oven 30 minutes, or until mixture is set and golden on top. Cool slightly before serving. Serves 6.

Parisian Chicken-Vegetable Ragout

The French have a particular talent for cooking poultry or game in a large pot with a medley of vegetables and agreeable seasonings so that the flavors are well preserved. This recipe is typical of numerous rich French stews called ragout.

6 tablespoons butter or margarine	1 cup chicken broth or water
2 frying chickens, about 3 pounds each, cut up	1 teaspoon dried rosemary
	Salt and pepper to taste
2 medium-sized onions, peeled and sliced	1½ cups fresh or frozen green peas
2 garlic cloves, crushed	1½ cups fresh or frozen cut-up green beans
1 can (1 pound, 12 ounces) tomatoes, undrained	

Melt butter in a large kettle. Wipe dry chicken pieces, and brown in hot melted butter on all sides, adding a few pieces at a time and removing them to a platter when done. Add onions and garlic to drippings, and sauté until tender. Return chicken to kettle. Add tomatoes, broth, rosemary, salt and pepper. Lower heat and cook slowly, covered, for 30 minutes. Add peas and beans, and cook about 10 minutes longer or until ingredients are tender—less time is needed for frozen than fresh vegetables. Serves 6 to 8.

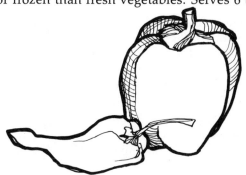

Especially for Company

An important aspect of entertaining is to offer good fare that usually includes one or more culinary specialties.

Flavorful and inviting vegetable dishes are excellent for company meals, as they are unusual, can often be prepared beforehand, and conform to the modern style of entertaining since they proffer simple elegance and economy rather than extravagance.

Consider, for example, such varied egg and vegetable dishes as omelets, soufflés, and baked combinations that find particular favor at brunch and luncheon, and vegetable casseroles, quiches, pies, and tarts that can be served as first or main courses at supper and dinner.

There can even be a great deal of pleasure in serving a single vegetable, such as a bunch of asparagus or celery or a whole cauliflower, cooked to perfection and elegantly garnished, either as a separate course or an accompaniment. The humble potato takes on a new aura when presented in a curry or in pancakes. Molded vegetable dishes are always bound to evoke compliments.

The recipes in this section have been chosen to display the versatility of vegetable cookery for company gatherings.

Persian Green-Vegetable Casserole

A nutritious and eye-appealing entrée for the main course at a luncheon or supper or as an accompaniment to beef or poultry for dinner.

- 1 cup finely chopped leaf lettuce
- 1 cup minced fresh parsley
- 2 cups chopped fresh spinach
- 2 cups finely chopped scallions, with tops
- 1½ tablespoons all-purpose flour
- ½ cup chopped walnuts
- Salt and pepper to taste
- 6 eggs, well beaten
- ¼ cup melted butter or margarine

Combine first six ingredients in a bowl. Season with salt and pepper. Add eggs and mix well. Pour butter into a 1½-quart baking dish. Add vegetable-egg mixture and spread evenly. Bake in preheated 350° F. oven about 45 minutes, or until mixture is set and a tester inserted into it comes out clean. Serve hot with plain yogurt as a sauce, if desired. Serves 6 to 8.

Omelette Basque

The Basque cooks of southern France and northern Spain have created an admirable selection of colorful and well-seasoned dishes, many of which feature eggs and vegetables. This is one of their best, good to serve at brunch or luncheon.

- ¼ cup olive or vegetable oil
- 2 large onions, peeled and sliced
- 1 to 2 garlic cloves, crushed
- 4 medium-sized tomatoes, peeled and chopped
- 2 large green peppers, seeded and chopped
- 2 canned pimientos, drained and chopped
- ½ teaspoon dried basil
- Salt and pepper to taste
- 8 eggs
- ¼ cup chopped fresh parsley

Heat oil in a large skillet. Add onions and garlic, and sauté until tender. Mix in vegetables, basil, salt and pepper, and cook 5 minutes. Add eggs, two at a time, stirring after each addition, and cook until mixture is set but softly scrambled. Stir in parsley just before cooking is finished. Serves 4.

Russian Potato Roll

An attractive and nourishing specialty for a winter dinner party. Serve with a beef entrée.

- 8 medium-sized potatoes
- Boiling salted water
- 2 eggs, well beaten
- 9 tablespoons butter or margarine
- 2 tablespoons chopped fresh parsley
- Salt and pepper to taste
- 2 cups cooked green peas
- 1 cup diced fresh or canned mushrooms
- 1/2 cup diced cooked carrots

Wash, peel, and boil potatoes in boiling salted water until just tender. While still warm, mash well. Add eggs, 6 tablespoons butter, and parsley. Season with salt and pepper. Beat until smooth. Turn out onto a flat surface and pat down to form a long narrow rectangle about 1 inch thick. Combine peas, mushrooms, and carrots. Add 2 tablespoons butter, and season with salt and pepper. Spoon lengthwise along center of potato rectangle. Fold potatoes around vegetables to enclose them and form a roll. Spread top with remaining 1 tablespoon butter. Place in a shallow rectangular baking dish. Bake in preheated 450° F. oven about 15 minutes, until golden on top. To serve, cut into thick slices. Serves 8.

Chinese Egg Foo Yong

This well-known Chinese dish, made with eggs and vegetables as well as meat or fish, is an excellent main course for a brunch or luncheon.

2	tablespoons soy sauce	¼	cup diced cooked ham
1½	tablespoons cornstarch	½	cup sliced fresh mushrooms
1	cup beef bouillon or water	⅓	cup sliced celery
1	teaspoon sugar	⅓	cup sliced bamboo shoots
⅓	cup minced scallions	⅓	cup chopped green peppers
2	tablespoons peanut or vegetable oil	6	eggs
			Pepper to taste

Heat soy sauce in a small saucepan. Mix cornstarch into bouillon, and add, with sugar, to soy sauce. Mix well and heat, stirring, until thickened. Keep warm. Sauté scallions in heated oil in a large skillet. Add ham and vegetables, and sauté, stirring, for 1 minute. Beat eggs slightly and season with salt and pepper. Pour over vegetable mixture. Tilt pan to spread evenly and cook slowly until mixture is set and dry on top. Turn out onto a warm plate. Grease pan lightly and return pancake-like mixture to skillet. Cook until golden on other side. Cut into wedges and serve with warm sauce. Serves 4.

Mexican Colache

Mexicans are truly experts in the cookery of corn, and one of their best ways of using fresh corn is in this colorful vegetable dish that dates back to Aztec times.

1	medium-sized onion, peeled and chopped	1	cup cut-up green beans
			About 1 cup water
2	tablespoons butter or margarine	4	medium-sized ears of corn, shucked and cleaned
4	zucchini or other summer squash, ends trimmed and sliced	2	large tomatoes, peeled and chopped
			Salt and pepper to taste
1	green pepper, cleaned and chopped		

Sauté onion in heated butter in a saucepan until tender. Add squash, green pepper, green beans, and water. Cook slowly, covered, 5 minutes. Cut up corn into 1-inch pieces, and add, with tomatoes, salt and pepper, to vegetable mixture. Cook, covered until ingredients are tender, about 10 minutes. Add a little more water during cooking, if needed. Serves 6.

Swiss Onion Tart

The Swiss prepare some of the world's most notable onion dishes, such as this tart, which makes an excellent entrée for a weekend midday meal.

- 2 thin slices bacon, chopped
- 2 large yellow onions, peeled and thinly sliced
- 3 tablespoons butter or margarine
- 2 eggs, lightly beaten
- ½ cup heavy cream
- 1 teaspoon sugar
- Freshly grated nutmeg to taste
- ⅓ cup grated Parmesan or Swiss cheese
- Salt and pepper to taste
- 1 9-inch baked pastry shell

Fry bacon until crisp; drain on absorbent paper and set aside. Sauté onions in heated butter in a skillet until tender, being careful not to brown. Combine remaining ingredients in a bowl, except for pastry shell. Add onions and mix well. Turn into pastry shell. Spread evenly and top with cooked bacon. Bake in a preheated 375° F. oven until set, about 35 minutes. Serve warm. Serves 4 to 6.

Algerian Baked Eggs With Vegetables

This traditional dish, called *chakchouka*, is an appealing egg-vegetable combination that may be served at brunch, at luncheon, or for a late supper.

- 1 large onion, peeled and chopped
- 2 garlic cloves, crushed
- 3 tablespoons olive or vegetable oil
- 2 medium-sized green peppers, cleaned and sliced
- 3 medium-sized tomatoes, peeled and sliced
- 1 small hot red pepper, cleaned and chopped, or 1 teaspoon ground red pepper
- Salt and pepper to taste
- 6 eggs
- 3 tablespoons chopped fresh parsley

Sauté onion and garlic in oil in a skillet until tender. Add vegetables, salt and pepper, and sauté until tender, 2 to 3 minutes. Spoon into a round shallow baking dish. Beat eggs in a bowl. Season with salt and pepper. Pour over vegetables, and tilt dish to spread evenly. Sprinkle with parsley. Bake in preheated 350° F. oven until set, about 20 minutes. Serves 4.

Near Eastern Potato-Cauliflower Curry

Two white-colored vegetables are combined in this curry to make an interesting main dish to serve at a late-evening supper or as an accompaniment to lamb.

3 tablespoons butter or margarine	3 large tomatoes, peeled and chopped
2 tablespoons curry powder	1 medium-sized cauliflower, cleaned, divided into florets
2 teaspoons ground turmeric	4 medium-sized potatoes, peeled and cubed
½ teaspoon paprika	
Salt and pepper to taste	About 1 cup water
1 large onion, peeled and chopped	1 tablespoon fresh lemon juice

Melt butter in a saucepan. Add spices, salt, and onion, and cook, stirring, 1 minute. Add remaining ingredients and cook slowly, covered, until vegetables are tender, about 20 minutes. Add more water during cooking, if needed. Serves 4.

Flemish Asparagus

Europeans are devoted to their tender pearl-white asparagus, grown underground and carefully harvested by hand each spring. In season only a short time, it is prized as the "king of vegetables" and is carefully cooked to preserve the nutrients and appearance. The best way to eat asparagus is with the fingers, picking up each stalk by the thick end. It makes an excellent luncheon dish that can be served with boiled new potatoes and thick slices of cooked bacon or ham, if desired. White asparagus is sold fresh in some American markets but is widely available in cans or jars.

>2 pounds fresh white or green asparagus
>Boiling salted water
>6 eggs
>¾ cup melted butter
>¾ cup chopped fresh parsley
>Salt and freshly ground black pepper to taste

Wash asparagus under running water. Cut off any tough stem ends and remove any large scales. Arrange in a large skillet and add ½ inch boiling salted water. Cook, covered, over moderate heat until tender, 10 to 12 minutes. Carefully remove from pan with tongs and drain. Place on a warm platter. Meanwhile, cook eggs in shells about 5 minutes, until whites are hard but yolks a little runny. Shell and chop. Mix with butter and parsley. Season with salt and pepper and spoon over asparagus. Serves 4.

Note: Canned asparagus can be used as a substitute for the fresh if desired.

Indian Vegetable Pilau

This colorful vegetable-rice dish may be served either as an entrée or as an accompaniment to roasted poultry or meat.

>1 large onion, peeled and chopped
>1 or 2 garlic cloves, crushed
>3 tablespoons butter or margarine
>2 teaspoons ground turmeric
>½ teaspoon cayenne
>½ teaspoon ground ginger
>½ teaspoon ground cumin
>Salt and pepper to taste
>3 medium-sized tomatoes, peeled and chopped
>1 cup cooked cut-up green beans
>1 cup diced cooked carrots
>1 cup cooked green peas
>2 cups cooked long-grain rice
>1 cup plain yogurt at room temperature

Sauté onion and garlic in butter in a skillet until tender. Add spices and salt and pepper, and cook 1 minute. Mix in vegetables and sauté until hot. Add rice and mix well. Leave on stove until hot and mix in yogurt. Cook another minute. Serves 4 to 6.

Spinach Ring with Creamed Eggs and Walnuts

Serve at a women's luncheon or weekend brunch.

Spinach Ring
- 1 tablespoon butter or margarine
- 1 tablespoon all-purpose flour
- 1 cup milk
- 1/4 teaspoon grated nutmeg
- Salt and pepper to taste
- 3 cups chopped cooked spinach
- 3 eggs, beaten

Melt butter in a medium-sized saucepan. Add flour and cook 1 minute. Gradually add milk, stirring, and cook until smooth and thickened. Season with nutmeg, salt and pepper. Mix in spinach and eggs, and turn into a greased ring mold. Put in a pan with 1 inch of water and bake in preheated 350° F. oven until set, about 45 minutes. Unmold on a warm platter. Fill with Creamed Eggs and Walnuts (see below) and serve at once. Serves 8.

Creamed Eggs and Walnuts
- 8 hard-cooked eggs, shelled
- 1 can (10 1/2 ounces) condensed mushroom soup
- 1/4 cup grated Swiss or Parmesan cheese
- 1/2 cup chopped walnuts

Cut eggs into quarters. Put remaining ingredients into saucepan and cook until hot. Add eggs and cook long enough for them to heat through. Spoon into the center of the Spinach Ring.

Albanian Minted Vegetable Casserole

An excellent dish whether served as the main dish at luncheon or supper or as an accompaniment to lamb or beef.

3 tablespoons butter or margarine	1 cup green peas
3 medium-sized potatoes, peeled and sliced thinly	½ cup tomato paste
	1 cup water
3 medium-sized tomatoes, peeled and sliced	½ cup olive or vegetable oil
	Salt and pepper to taste
2 medium-sized zucchini, stems removed and sliced	3 tablespoons chopped fresh mint

Heat butter in a skillet and fry potatoes in it until soft. Arrange potatoes, tomatoes, zucchini and peas in layers in a greased casserole. Combine tomato paste and water in a small saucepan. Bring to a boil. Stir in oil, salt and pepper. Add mint. Pour over vegetables. Bake in preheated 350° F. oven until vegetables are tender, about 30 minutes. Serve 4 to 6.

Chinese Vegetable Fried Rice

Although this Oriental specialty is very well known, it is always welcome. The combination of cooked rice, crisp vegetables, and piquant seasonings is superb. A fine dish for brunch.

½ cup chopped scallions, with tops	½ cup shelled fresh or defrosted frozen peas
3 tablespoons peanut or vegetable oil	4 cups chilled cooked rice
	2 tablespoons soy sauce
½ cup sliced fresh or canned mushrooms	½ teaspoon sugar
	Pepper to taste
¼ cup sliced bamboo shoots	2 eggs, lightly beaten
½ cup bean sprouts	

Sauté scallions in heated oil in a wok or skillet until tender. Add vegetables and stir-fry until peas are just tender. Add rice and stir-fry until hot. Mix in soy sauce and sugar. Season with pepper. Pour in eggs and cook, stirring, until set and hot. Serve at once. Serves 4.

Red and Green Pepper Soufflé

A light and flavorful soufflé for a special celebration or holiday dinner. Easy-to-prepare and a delight to serve.

About 6 tablespoons butter or margarine
2 tablespoons fine dry bread crumbs
2 tablespoons grated Parmesan cheese
2 medium-sized green peppers, cleaned and chopped
2 medium-sized sweet red peppers, cleaned and chopped
¼ cup all-purpose flour
1 cup light cream or milk
4 eggs, separated
¼ teaspoon dried marjoram
Salt and pepper to taste

Grease well sides and bottom of a 6-cup soufflé dish with 1 tablespoon butter. Combine crumbs and cheese and sprinkle over sides and bottom, tilting to spread evenly. Heat 2 tablespoons of butter in a skillet, add peppers, and sauté until soft; set aside. Melt 3 tablespoons butter in a saucepan and add flour. Cook, stirring, over low heat for 1 minute. Gradually add cream and cook, stirring, over low heat until thickened and smooth. Remove from heat and beat in egg yolks one at a time. Add sautéed peppers, marjoram, salt and pepper. Cook, stirring, for a minute or two. Cool slightly. Beat egg whites until stiff and fold gradually into mixture. Pour into soufflé dish and bake in preheated 375° F. oven until set, about 25 minutes. Serves 4 to 6.

Iraqi Vegetable Pie

This colorful vegetable pie made with a potato "pastry" can be served as a unique specialty at luncheon or as an accompaniment at dinner to lamb or beef.

8 medium-sized potatoes Boiling salted water 6 tablespoons butter or margarine Salt and pepper to taste 6 scallions, with tops, cleaned and minced	2 medium-sized cucumbers 4 large tomatoes, peeled and sliced 2 tablespoons sugar

Scrub potatoes and cook in boiling salted water to cover until tender. Drain; remove skins and mash while still warm. Add butter, salt and pepper, and scallions. Mix well. Spoon into a shallow greased baking dish. Spread evenly. Wash cucumbers and remove ends. Line lengthwise with tines of a fork and slice thinly. Arrange alternately with tomato slices in rows or circles, overlapping each other, over potatoes. Sprinkle with sugar. Bake in preheated 350° F. oven 30 minutes. Serve in baking dish. Serves 8.

Italian Macaroni-Zucchini Timballo

This easy-to-prepare version of the traditional Italian *timballo*, a filled "pie" baked in a round form, is equally good served as an accompaniment to meat or as a luncheon specialty.

2 pie-crust sticks 2 cups elbow macaroni Boiling salted water ½ cup chopped scallions, with tops	2 tablespoons butter or margarine 2 cans (1 pound each) zucchini in tomato sauce ½ cup grated Parmesan cheese

Prepare 1 pie-crust stick according to package directions. Roll out and line a 9- by 1½-inch round cake pan with pastry. Cook macaroni in boiling salted water until tender; drain. Sauté scallions in heated butter in a saucepan until tender. Add zucchini and heat. Stir in cooked and drained macaroni and spoon into pastry-lined cake pan. Sprinkle top with cheese. Prepare and roll out remaining pie-crust stick. Fit over top of dish. Make several slits in pastry. Bake in preheated 375° F. oven until pastry is crisp and golden, 30 to 40 minutes. To serve, cut into wedges. Serves 6 to 8.

Fresh Mushroom Soufflé

An elegant luncheon dish for a women's get-together.

- 4 eggs at room temperature
- 1 additional egg white
- 1 tablespoon minced shallots or scallions
- 4 tablespoons butter or margarine
- 1 tablespoon fresh lemon juice
- 1 cup finely chopped fresh mushrooms
- Dash freshly grated nutmeg
- Salt and pepper to taste
- 3 tablespoons all-purpose flour
- 1 cup light cream or milk

Separate eggs, putting yolks in a small bowl and whites in a large one. Sauté shallots, or scallions, in 1 tablespoon heated butter with lemon juice in a medium-sized saucepan until tender. Add mushrooms and sauté 3 minutes. Add seasonings and remove from heat. Melt remaining 3 tablespoons butter in a saucepan. Stir in flour and cook, stirring, 1 minute. Gradually add cream or milk, stirring constantly, and cook slowly until smooth and thick. Remove from heat and cool slightly. Beat egg yolks until lemon-colored and creamy. Mix some of the hot sauce with the yolks and return all to pan. Stir in mushroom mixture. Beat egg whites until stiff. Carefully fold half of them into mushroom mixture. Then add remaining half. Spoon mixture into a buttered 1½-quart soufflé dish. Bake, uncovered, in preheated 375° F. oven until golden and puffed up, about 35 minutes. Serve at once. Serves 4.

Crookneck Squash-Green Bean Quiche

The familiar summer squash, called crookneck because of its curved neck, has a warted yellow skin and is best when about half mature. It combines well with green beans to make a vibrant quiche, good for a brunch or luncheon.

⅓ cup chopped scallions, with tops	4 eggs
2 tablespoons butter or margarine	1 teaspoon sugar
	½ teaspoon dried marjoram
2 cups sliced fresh or 1 package (10 ounces) frozen sliced crookneck squash, defrosted	½ cup plain yogurt or sour cream
	Salt and pepper to taste
	1 deep-dish 9-inch pastry shell, baked 10 minutes and cooled
1 cup cut-up fresh or ½ package (9 ounces) frozen Italian green beans, defrosted	

Sauté scallions in heated butter in a skillet until tender. Add squash and beans, and sauté until just tender. Combine eggs, sugar, marjoram, yogurt, and salt and pepper in a bowl. Add sautéed vegetables. Turn into pie shell, place in middle level of preheated 375° F. oven, and bake about 35 minutes, until puffed and golden and a knife inserted into the center comes out clean. Serves 6.

California Spinach-Zucchini Casserole

This is an inviting vegetable casserole for a dinner or buffet.

2 packages (10 ounces each) frozen chopped spinach	¼ teaspoon dried basil
	Salt and pepper to taste
3 slices white bread, crusts removed	1 cup sliced scallions, with tops
	About ¼ cup olive or vegetable oil
½ cup light cream or milk	
4 eggs, slightly beaten	4 medium-sized zucchini, stems trimmed and sliced
¼ cup grated Parmesan cheese	
½ cup chopped fresh parsley	

Cook spinach until tender, and drain, pressing with a large spoon to release liquid. Turn spinach into a large bowl. Soak bread in cream or milk until soft. Mash with a fork and add to spinach. Add eggs, cheese, parsley, basil, salt and pepper. Sauté scallions in heated oil in a skillet until tender. Add zucchini slices and cook slowly, covered, turning occasionally until just tender. Add to spinach mixture. Spoon into a 12- by 8- by 2-inch greased baking dish. Bake in preheated 350° F. oven until set, about 45 minutes. Serves 12.

Summer Broccoli Ring with Sour Cream Sauce

An excellent luncheon or supper vegetable specialty.

1½ tablespoons butter or margarine	1 package (10 ounces) frozen chopped broccoli, defrosted
1½ tablespoons all-purpose flour	¼ cup grated Parmesan cheese
1½ cups milk	1½ cups sour cream
¼ teaspoon dried oregano	2 teaspoons minced chives
Salt and pepper to taste	1 tablespoon chopped fresh dill
4 eggs, beaten	

Melt butter in a medium-sized saucepan. Blend in flour and cook 1 minute. Gradually add milk and cook slowly, stirring, until thick and smooth. Season with oregano, salt and pepper. Mix some of the hot sauce with the eggs and return all to saucepan. Cook 1 minute and remove from heat. Add broccoli and cheese and mix well. Turn into a greased ring mold. Place in a pan with hot water in preheated 350° F. oven and bake until cooked and set, about 45 minutes. Unmold on a warm platter. Meanwhile, mix sour cream and chives together in a saucepan and cook until heated. Pour into center of broccoli ring and garnish with dill. Serves 8.

Pennsylvania-Dutch Potato Pancakes

An excellent company dish to serve with roast pork, baked ham, or pork chops.

2 cups seasoned mashed potatoes	1 cup milk
2 eggs, separated	1 small onion, peeled and minced
¼ cup all-purpose flour	Fat for frying
2 teaspoons baking powder	
Salt and pepper to taste	

Combine potatoes, egg yolks, flour, baking powder, and salt and pepper in a bowl. Add milk and onion, and mix until smooth. Beat egg whites until stiff, and fold into potato mixture. Drop by tablespoonfuls into a well-greased skillet. Brown on both sides, adding more fat if needed. Serve at once. Serves 4 to 6.

New Mexican Chiles Rellenos

Stuffed peppers, a Mexican specialty that is very popular in the Southwest, are usually made with a filling of cheese although sometimes beans are used. A good dish for company because the flavor is exceptional, and the peppers look most attractive.

1 can (7 ounces) whole green chiles	4 eggs, separated
½ pound Monterey Jack or Cheddar cheese	1 tablespoon water
About ¾ cup all-purpose flour	Salt and pepper to taste
	Fat or oil for frying
	Chili sauce (optional)
	Grated cheese (optional)

Drain chiles. Rinse and cut a slit down the side of each. Remove seeds and membranes. Stuff each chili with a piece of cheese cut about ½ inch wide, ½ inch thick, and 1 inch shorter than the chili. Fold over edges of chili to cover cheese. Roll chilis in flour to coat on all sides. Beat egg whites until stiff. Mix egg yolks, previously beaten, with 4 tablespoons flour, the water, and a little salt. Fold into whites. In a greased omelet pan or skillet put ½ cup egg mixture, heaping it into a mound. Immediately put a stuffed chili in the center of the mound and cover with some of mixture to enclose chili. Cook about 3 minutes. Turn and cook until golden, about 3 minutes longer. Serve, if desired, with chili sauce and grated cheese. Serves 4.

Celery Victor

This famous salad, originated by a San Francisco chef at the St. Francis Hotel, is unusually attractive. Serve at a summer luncheon or dinner party.

4 celery hearts	2 hard-cooked eggs, shelled and sliced
1½ cups chicken broth	
1 bouquet garni (1 bay leaf, 2 parsley sprigs, ¼ teaspoon dried thyme tied together in cheesecloth)	2 medium-sized tomatoes, peeled and sliced
	6 flat anchovy fillets, chopped
	3 tablespoons chopped fresh parsley
Salt and pepper to taste	
½ cup French dressing	

Split each celery heart in half lengthwise. Cut out root and trim leaves, leaving only baby ones. Wash celery to remove all dirt. Arrange in a shallow flameproof dish or skillet in which celery can lie flat. Cover with chicken broth. Add *bouquet garni,* salt and pepper. Bring to a boil. Reduce heat and cook slowly, covered, until just tender, about 12 minutes. Cool in broth. Drain celery and arrange on a platter. Top with French dressing and garnish with egg and tomato slices. Sprinkle with anchovy fillets, parsley, and pepper. Serves 4.

Garnished Whole Cauliflower from Sicily

One of the most attractive and appealing ways of serving cauliflower is the whole head adorned with a colorful sauce.

1 whole cauliflower	1 cup tomato sauce
Salted water	½ teaspoon dried basil
1 tablespoon fresh lemon juice	1 teaspoon sugar
1 large onion, peeled and chopped	Salt and pepper to taste
	¼ cup chopped fresh parsley
1 to 2 garlic cloves, crushed	6 flat anchovy fillets
2 tablespoons olive or vegetable oil	

Cut off base and tough outer leaves of cauliflower. Wash in cold running water, holding upside down. Heat 1 inch salted water in a large kettle; add lemon juice. Cook cauliflower, uncovered, 5 minutes. Cover and boil until just tender, about 20 minutes. Drain; arrange on a hot vegetable dish and keep hot. While cauliflower is cooking, prepare sauce. Sauté onion and garlic in heated oil in a skillet. Add tomato sauce, basil, sugar, salt and pepper. Cook slowly, uncovered, 10 minutes. Add parsley. Spoon over hot cooked cauliflower and garnish with anchovies. Serves 4 to 6.

Greek Spinach Pie

This flaky large pie called *spanakopitta*, made with thin sheets of pastry sold as *phyllo* or *fila* in specialty food stores, is excellent for a women's luncheon. The pie is light, tasty, and attractive.

- 1 package (10 ounces) frozen spinach
- ½ cup chopped scallions, with tops
- 1 tablespoon olive oil or butter
- 2 cups small-curd cottage cheese
- 1 cup crumbled Feta cheese or grated Muenster cheese
- ¼ cup chopped fresh parsley
- 2 tablespoons chopped fresh dill
- Salt and pepper to taste
- ½ pound phyllo sheets
- About ¾ cup melted butter or margarine

Defrost spinach and chop. Pour off any liquid. Put in a large bowl. Sauté scallions in heated oil until tender. Add spinach, cottage cheese, Feta or Muenster cheese, parsley, dill, salt and pepper, and mix well. Grease a shallow oblong 12- by 9-inch baking dish with butter. Place one third of pastry sheets in dish, having first brushed each one with melted butter and folded it over in half. Spread half of spinach mixture over the pastry sheets in the dish. Repeat with another third of sheets, another layer of spinach mixture, and remaining third of sheets, brushing each sheet with melted butter. Brush top generously with butter. Bake in preheated 350° F. oven until golden and crisp on top, about 40 minutes. Cut into squares and serve warm. Serves 6 to 8.

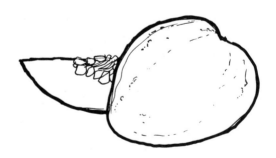

Greens & Leafy Vegetables

From the standpoint of health, greens and leafy vegetables are the most valuable of all. There is a wide variety of wild and cultivated types, and even those commonly thought of as weeds—like the dandelion—can be made into delicious, highly nutritious, and attractive dishes.

Many of our American greens and leafy vegetables were originally introduced by early settlers from other lands, where they had long been highly prized. Because fresh green vegetables were not available during the winter, the spring greens were especially welcome and also served as a healthful tonic. Recipes calling for "salat greens" in old cookery books were intended for wild greens and potherbs.

It is still a springtime custom in some parts of the country to pick wild greens. They must, however, be selected with care while still young and tender, or the cooked dish will be harsh and unappealing. Among the most familiar wild greens are dandelion, purslane, lamb's-quarters, pigweed, wild mustard, sorrel, dock, watercress, burdock, chickory, poke stalk, milkweed, and fiddlehead fern. They are all generally cooked in the same way—simmered in a little water with fat or other flavoring.

Southerners became particularly fond of a number of

greens, such as collards, kale, mustard and turnip greens, that are still staple fare, and are seasoned with ham or bacon, vinegar, lemon juice, brown sugar, onions, mustard, nutmeg, dill, caraway seed, or celery seed. Other vegetables cooked as greens are spinach, beet and radish tops, Swiss chard, and corn salad.

All greens should be carefully picked over; any imperfect leaves should be discarded, and the rest washed in water several times. Most greens can be cooked using only the water that clings to the leaves after washing. Drain well after cooking, but save the liquid for later use in soups or gravies; it can also be drunk. Season the greens according to taste.

Many greens are also used uncooked, particularly in salads. Lettuce is the most important of the salad greens, and four types are generally sold: iceberg, butterhead, romaine, and leaf. Escarole and chicory or curly endive are also widely available. Although they are not usually cooked, these greens can be steamed, braised, or baked.

The wide assortment of these vegetables deserves to be better known and more frequently used, for they contain important minerals and vitamins—especially iron, calcium, and riboflavin—are a most important source of vitamin A, and are very low in calories.

While the recipes in this section contain a number of examples from foreign lands, a great many of them are for native American dishes.

Scotch Kale and Oatmeal

Kail, as the Scots call their treasured kale, is sometimes the only green vegetable that they have to eat during the long cold winter. Generally it is boiled and seasoned with butter, salt and pepper, or combined with mashed boiled potatoes, onions, and milk. Here is another typical recipe.

½ cup chopped scallions, with tops	Salt and pepper to taste
3 tablespoons bacon drippings or other fat	1 cup cooked oatmeal
	¼ cup milk
2 pounds fresh kale, washed, stemmed, and chopped	Dash grated nutmeg

The Delectable Vegetable

Sauté scallions in heated bacon drippings in a large saucepan until tender. Add kale and about ¾ cup water and cook, covered, until kale is tender, about 12 minutes. Drain and mix with salt, pepper, oatmeal, milk, and nutmeg. Leave on the stove long enough to heat through. Serves 4 to 6.

Green-Leaf Curry from India

Another interesting version of the ancient dish of curry that is rich in both vitamins and flavor.

2 tablespoons vegetable oil	3 cups chopped cleaned mustard greens
1 tablespoon turmeric powder	
½ teaspoon ground red pepper	3 cups chopped cleaned romaine or leaf lettuce
1 teaspoon ground coriander	
2 garlic cloves, crushed	⅓ cup beef bouillon or water
2 medium-sized onions, peeled and sliced	Salt and pepper to taste
	Juice of 1 lemon
3 cups chopped cleaned spinach	

Heat oil in a large skillet. Add spices and cook 1 minute. Mix in garlic and onion, and sauté until tender. Add greens and bouillon, and cook slowly, stirring lightly now and then, until greens are wilted and tender, about 8 minutes. Season with salt and pepper, and add lemon juice. Serves 4 to 6.

Early American Wilted Lettuce

An old and treasured American salad that was very popular with pioneer families who often prepared it with wild greens and salt pork.

2 small heads leaf lettuce, washed and drained	1 tablespoon chopped fresh dill or parsley
4 slices thin bacon, chopped	Salt and pepper to taste
½ cup chopped scallions	1 hard-cooked egg, shelled and chopped
¼ cup vinegar	
1 tablespoon sugar	

Tear lettuce into bite-size pieces and put in a salad bowl. Fry bacon until crisp. Remove from pan and drain on absorbent paper. Add scallions to heated fat and sauté until tender. Stir in vinegar and sugar and bring to a boil. Add, with dill, salt and pepper, to lettuce and toss. Garnish with crisp bacon bits and chopped egg. Serves 4.

Mormon Pioneer Greens

While the dedicated Mormons pioneered across our country to Utah, they relied on edible wild greens for nutrients. Later their cooks created a number of good dishes made with both wild and cultivated greens, and this recipe is typical of them.

6 thin slices bacon, chopped	1 egg, slightly beaten
1 small onion, peeled and minced	3 tablespoons water
	⅓ cup vinegar
2 tablespoons all-purpose flour	1 cup light cream or milk
2 tablespoons sugar	3 cups chopped cooked greens (spinach, mustard, turnip, or dandelion)
⅛ teaspoon salt	
¼ teaspoon powdered mustard	

Fry bacon in a medium-sized saucepan until crisp. Add onion and sauté until tender. Stir in flour and cook 1 minute. Add sugar, salt, mustard, egg, water, and vinegar, previously mixed together, and cook slowly, stirring, until mixture is thick and smooth. Gradually add cream or milk and continue cooking about 1 minute. Stir in greens and leave on stove long enough to heat through. Serves 4.

Spinach, Roman Style

The Romans are very fond of fresh spinach. They cook it with anchovies and lemon juice, use it in croquettes, puddings, and dumplings as well as in salads. This recipe is a favorite Roman specialty.

1	package (10 ounces) fresh spinach	1	or 2 garlic cloves, crushed
1½	teaspoons salt	⅓	cup pine nuts
2	tablespoons butter	⅓	cup seedless raisins, previously softened in water 5 minutes
1	tablespoon olive oil		
1	small onion, peeled and minced		Pepper to taste

Wash and trim spinach leaves, cutting off stems and discarding any wilted leaves. Put in a deep saucepan with whatever water remains on the leaves. Season with salt. Cook slowly, covered, until tender, about 12 minutes. Drain well, pressing with a spoon to release all liquid; chop coarsely. Heat butter and oil in a skillet and sauté onion until tender. Mix in spinach and garlic and sauté, turning several times, until hot. Add nuts, raisins, and pepper, and cook over low heat, stirring, about 5 minutes. Serves 4.

Wilted Dandelion Greens

Young tender dandelion greens, freshly picked, were once welcome spring greens, treasured for their bitter taste and as a difference from the winter root vegetables; they were also considered a valuable spring tonic. Generally they were used in salads or cooked in this way. Dandelions should be picked before they blossom. Good seasonings for wild or cultivated dandelions are bacon, mint, garlic, vinegar, and mustard.

2 quarts dandelion greens	1 teaspoon salt
¼ cup bacon fat	2 tablespoons chopped fresh parsley
¼ cup cider vinegar	
2 teaspoons sugar	

Wash greens thoroughly and cut off any stems. Cut into small pieces with scissors. Heat bacon fat, vinegar, sugar, and salt in a heavy saucepan. Wipe greens dry and add. Cook over moderate heat, covered, until wilted and tender, about 12 minutes. Stir in parsley. Serves 4 to 6.

French Chicory Gratinée

Chicory, also called curly endive, has feathery leaves with dark green edges and white centers. Because of its slightly bitter taste, chicory is usually combined with other greens when it is used in salad; it can, however, also be cooked, when it is often braised or, as in this dish, combined with a cream sauce.

1 pound fresh chicory	Dash grated nutmeg
1 cup water	Salt and pepper to taste
4 tablespoons butter or margarine	1 cup cooked or canned small white onions
4 tablespoons all-purpose flour	½ cup fine dry bread crumbs
1½ cups light cream or milk	½ cup grated Parmesan cheese
1 teaspoon sugar	

Cut off stem ends from chicory and discard any wilted leaves. Wash thoroughly to remove dirt. Put in a saucepan with water and cook, covered, over medium heat until tender, about 12 minutes. Drain and chop. Melt 3 tablespoons butter in a saucepan, and mix in flour. Cook 1 minute. Gradually add cream or milk and cook slowly, stirring, until thick and smooth. Stir in sugar, nutmeg, salt, pepper, onions, and chopped chicory. Spoon into a shallow baking dish. Sprinkle with bread crumbs and cheese, and dot with remaining 1 tablespoon butter. Bake in preheated 375° F. oven until hot and bubbly, about 25 minutes. Serves 4.

Collards with Sour-Cream Dressing

A member of the cabbage family and closely related to kale, collards are grown extensively in our South. The long smooth leaves are an excellent source of vitamin A, vitamin C, iron, and calcium.

2 pounds fresh collards	1 tablespoon lemon juice
1/4 cup water	About 1/2 cup sour cream
1/2 teaspoon salt	1 hard-cooked egg, shelled and chopped
4 slices thin bacon, chopped	

Trim stems off collards and wash carefully to remove sand. Put in a large saucepan with water and salt. Cook slowly, covered, until tender, about 12 minutes. Meanwhile fry bacon until it is crisp and the fat is rendered. Drain bacon and set aside. Combine 2 tablespoons bacon fat, lemon juice, and sour cream, and add to collards. Leave on stove long enough to heat through. Garnish with chopped egg and bacon. Serves 4 to 6.

Belgian Creamed Beet Greens

The Belgians have long appreciated fresh young beet greens. In fact, some like the beet tops more than the tubers. Beets are related to chard and thrive in a cool climate. Some beet greens are sold separately, while others can be cut from bunches of beets, but they should always be young and tender.

4 slices thin bacon, chopped	1/2 cup light cream or milk
1 medium-sized onion, peeled and chopped	1/4 cup prepared horseradish
1 quart chopped washed young beet greens	1 teaspoon sugar
1/4 cup beef bouillon or water	1/2 teaspoon powdered mustard
	Salt and pepper to taste

Fry bacon in a large saucepan to render fat. Drain off all but 2 tablespoons of fat. Add beet greens and bouillon. Cook slowly, covered, until tender, about 12 minutes. Drain off any liquid and add remaining ingredients. Cook slowly, uncovered, 5 minutes. Serves 4.

Baked Mustard Greens and Cottage Cheese

A different dish of greens that is nutritious as well as appealing.

- 1 package (10 ounces) frozen chopped mustard greens
- 1 cup cottage cheese
- 1 small onion, peeled and minced
- 1/2 cup finely chopped scraped carrots
- 1/2 cup ketchup
- 1 tablespoon Worcestershire sauce
- Salt and pepper to taste

Cook mustard greens according to package directions. Drain well. Combine in a bowl with remaining ingredients, and spoon into a shallow baking dish. Bake in preheated 350° F. oven for 25 minutes until hot and bubbly. Serves 4.

Oriental Stir-Fry Greens

An excellent main dish or accompaniment.

- 1 package (10 ounces) fresh spinach, washed, trimmed, and cut up
- 4 cups cleaned chopped fresh mustard greens
- 1 bunch Chinese or celery cabbage, washed, trimmed, and cut up
- About 3 tablespoons peanut or vegetable oil
- 2 garlic cloves, crushed
- 2 large onions, peeled and sliced
- 1 cup sliced bamboo shoots
- 3 tablespoons soy sauce
- 3 tablespoons chicken broth or water
- 1 tablespoon sugar
- Pepper to taste

Wipe dry greens and cabbage. Heat oil in a large skillet or wok and add garlic and onions. Sauté until tender. Add greens and stir-fry 1 or 2 minutes. Then add remaining ingredients and stir-fry, turning carefully, about 5 minutes, until tender but a little crisp. Serves 6 to 8.

Butter-Steamed Lettuce with Cheese

2 small heads leaf lettuce
¼ cup butter or margarine
Salt and pepper to taste
½ cup grated Parmesan cheese

Wash, clean, and dry lettuce. Break into small pieces. Melt butter in a saucepan and add lettuce. Cook slowly, covered, until tender, about 3 minutes. Season with salt and pepper and add cheese. Serves 4.

French Puréed Sorrel

Sorrel, also called sour grass or dock, is an ancient green with an acid sour flavor that is used in soup and salad, to make a purée, or cooked with other vegetables. Although not widely available, sorrel can be found in some American markets and is valued for its high vitamin-A content.

3 pounds fresh sorrel
3 tablespoons butter or margarine
3 tablespoons all-purpose flour
1 cup light cream
1 teaspoon sugar
Salt and pepper to taste
2 egg yolks

Remove stems from sorrel and wash leaves well in warm water. Put in a large saucepan with a little water and cook, covered, until tender, about 8 minutes, stirring once or twice. Drain well. Purée or whirl in a blender. Melt butter in a saucepan and stir in flour. Cook 1 minute. Add cream, sugar, salt and pepper, and cook slowly, stirring, until thickened and smooth. Stir in sorrel. Mix some of hot sauce with egg yolks in a small bowl. Return mixture to saucepan and cook slowly, stirring, 1 or 2 minutes. Serves 4 to 6.

Puerto Rican Swiss Chard

A type of beet that does not develop fleshy roots, Swiss chard is grown for its large greenish leaves. Both stalk and leaves are eaten.

- 2 pounds fresh Swiss chard
- ¼ pound slab bacon, cut into 1-inch cubes
- 1 large onion, peeled and chopped
- 2 large tomatoes, peeled and chopped
- 1 large green pepper, cleaned and chopped
- ¼ cup beef bouillon or water
- Salt and pepper to taste

Wash chard and remove any stems. Break into small pieces, cutting out and discarding thick ribs. Fry bacon in a large saucepan to render fat. Add onion and sauté until tender. Mix in chard and remaining ingredients. Cook slowly, covered, until chard is tender, about 12 minutes. Serves 4 to 6.

Shaker Greens with Eggs

The innovative Shakers who worked hard and sincerely to improve their cookery were particularly fond of wild and cultivated greens, which they used in creative dishes like this one.

- 1 package (10 ounces) frozen chopped spinach
- 1 package (10 ounces) frozen chopped mustard greens
- 2 tablespoons butter or margarine
- ¼ cup chopped scallions, with tops
- 3 tablespoons chopped fresh parsley
- ¼ teaspoon dried rosemary
- Salt and pepper to taste
- ½ cup light cream or milk
- 6 poached eggs
- Grated nutmeg

Cook spinach and mustard greens according to package directions. Drain. Sauté scallions in butter in a saucepan. Add cooked spinach and greens, parsley, rosemary, salt and pepper, and mix well.

Add cream and cook slowly for 5 minutes to blend flavors. Turn into a serving dish and spread evenly. Spoon hot poached eggs over greens and top with nutmeg. Serves 6.

Poor Man's Asparagus

In Europe and early America scallions, also known as green or spring onions, were nicknamed "poor man's asparagus" as they were more available and less expensive than asparagus. Scallions are sweet and mild, and their tops may also be eaten.

- 4 bunches scallions
- 2 tablespoons butter or margarine
- 2 tablespoons all-purpose flour
- 1 cup light cream or milk
- Salt and pepper to taste
- 2 egg yolks
- Dash grated nutmeg
- 4 pieces of toast (optional)

Trim scallions, cutting off roots and any wilted tops, and leaving about 2 inches of green; wash well. Put in a heavy skillet and add a little water. Cook, covered, until tender, about 10 minutes. Meanwhile, melt butter in a saucepan and stir in flour. Gradually add cream or milk, and cook slowly, stirring steadily, until thick and smooth. Season with salt and pepper. Mix some of hot sauce with egg yolks. Return mixture to saucepan and cook until thick and smooth. Add nutmeg. Spoon over hot scallions, previously placed on toast, if desired. Serves 4.

Indonesian Mustard Greens

Chinese or celery cabbage and other greens may be cooked in this same way.

- 2 pounds fresh mustard greens
- 3 tablespoons vegetable oil
- 2 garlic cloves, crushed
- 1 large onion, peeled and sliced thinly
- 2 teaspoons minced red or green chilis
- 3 tablespoons soy sauce
- Salt and pepper to taste
- 1 tablespoon fresh lime or lemon juice

Wash greens well. Cut off stems and any wilted leaves. Cut up coarsely with scissors; wipe dry. Heat oil in a skillet and add garlic and onion. Sauté until tender. Add greens, chilis, soy sauce, salt and pepper. Cook slowly, covered, about 30 minutes. Mix in lime or lemon juice, and remove from heat. Serves 4.

Down-East Fiddleheads

Tender, green, fuzzy, and curled fern shoots, or sprouts, called fiddleheads (and also known as fern fronds, ostrich ferns, and cinnamon ferns) have long been treasured delicacies in northern Maine. Found growing on the shores of streams, they are picked when just pushing up through the ground in the spring, for they are at their best when young and tender. In yesteryear the greens provided a welcome variation from the lengthy monotonous winter diet of root vegetables. Today they are still sought-after spring fare.

Fresh fiddleheads must be thoroughly washed, and they then are steamed for a few minutes in a small amount of water. When cooked, they are generally eaten dressed only with butter, salt and pepper, but they are also delicious when served with Hollandaise or some other favorite sauce.

Fiddleheads also grow in other areas of northern New England, as well as in neighboring Canada, and they are available canned or frozen in some parts of the country.

Southern Collards

2 pounds fresh collards	Salt and pepper to taste
1/2 pound slab bacon, cut into 1/2 inch pieces	1/2 teaspoon Tabasco or hot red pepper sauce
1 large onion, peeled and sliced	2 tablespoons vinegar

Trim stems and remove any tough ribs from collards unless very young and tender. Wash well. Fry bacon in skillet until crisp; remove and drain. Pour off all but 2 tablespoons bacon fat. Sauté onion in it until tender. Add greens with water clinging to leaves and steam, covered, over low heat for 10 minutes. Stir in salt, pepper, red pepper sauce, and vinegar and cook another 5 minutes. Serves 4.

Vermont Mess o' Greens with Cornmeal Dumplings

An old and treasured country dish.

1/2 pound salt pork	6 medium-sized potatoes, peeled and halved
1 peck dandelion greens, cleaned and washed	Salt

Put salt pork in a heavy saucepan. Cover with water and cook over medium heat, covered, for 30 minutes. Add greens and potatoes, and continue cooking for 30 minutes. Take off cover and drop in dumplings (see below) by the spoonful over the greens. Cook, covered, for 15 minutes or until dumplings are tender. Serves 4.

Cornmeal Dumplings

1 cup sifted all-purpose flour	½ teaspoon salt
1 cup yellow cornmeal	1 egg
2 teaspoons baking powder	¾ cup milk

Sift flour, cornmeal, baking powder, and salt into a bowl. Combine egg and milk and mix with dry ingredients. Set aside until ready to cook. Mix again and drop into dandelion mixture above.

German Hop Sprouts

The scaly fruits of the hop plant, called hop sprouts, are primarily used in Europe and America to impart a characteristic lightness and bitterness to beer. In Germany and other countries in Europe the tender young hop sprouts are considered culinary delicacies and are cooked and eaten like asparagus. The fresh sprouts are only available for a short season, about the month of March when they burst from the plant.

Hop sprouts should be rapidly washed in water several times and then put in boiling salted water with lemon juice to cook until just tender. Once cooked, they can be served either hot, with melted butter or a cream sauce, or cold as an ingredient in salad.

Alabama Turnip Greens and Pot Likker

Southerners are very fond of fresh greens—turnip, mustard, or collard—cooked with salt pork and served with "pot likker," the liquid in which the greens are cooked. For this "likker," additional boiling water must be added to the greens while they are cooking. A favorite way of serving the "pot likker" is over cornbread squares.

1 large bunch, about 2 pounds, turnip greens	2 quarts water Salt to taste
½ pound salt pork	

Cut stems from greens and remove any wilted leaves. Wash well. Put salt pork and water in a kettle, and cook slowly, covered, for 1 hour. Add washed greens and cook slowly, covered, until tender, about 12 minutes. Do not overcook. Season with salt. Drain greens, reserving liquid. Slice pork and serve over cooked greens. Pour liquid over greens and pork. Serves 6.

Note: Another favorite way of preparing greens in the South is with hog jowl instead of salt pork.

Parisian Watercress Soup

A tangy aromatic herb, watercress belongs to the mustard family and is rich in calcium and vitamin C. The French truly appreciate watercress, which they use in sandwiches, salads, soups, and purées. It is welcomed with great fanfare in the spring, and there is even an old belief that watercress sharpens the memory. Watercress should be handled with care, washed gently in cold water, drained well, and used as soon as possible.

2 bunches watercress	4 tablespoons all-purpose flour
1 cup boiling salted water	Salt and pepper to taste
4 tablespoons butter or margarine	3 cups light cream or milk
1 small onion, peeled and minced	

Wash watercress and remove any stems. Put in a saucepan with the boiling salted water. Cook, covered, until tender, about 10 minutes. Press through a sieve, or purée. Heat butter in a saucepan. Add onion and cook until tender. Mix in flour, salt and pepper, and cook 1 minute. Gradually add cream or milk and cook until thick and smooth. Add watercress and liquid and leave on stove long enough to heat through. Serves 4 to 6.

Virginia Cress Stew

A dry-land cress, also called creecy greens, is grown in the fields of Virginia. Only there and in neighboring areas is it generally available fresh, but it is also sold in cans. Another name for the cress is peppergrass, which comes from its peppery flavor.

3 slices thin bacon, chopped	3 medium-sized potatoes, peeled and cubed
1 large onion, peeled and chopped	1 can (15 ounces) dry-land cress
3 stalks celery, cleaned and chopped	2 cups canned whole kernel corn
1 can (1 pound) tomatoes, undrained	½ teaspoon dried thyme
	Few drops Tabasco sauce

Fry bacon in a large saucepan until crisp. Pour off all fat except 2 tablespoons. Add onion and sauté until tender. Mix in celery, tomatoes, and potatoes. Cook slowly, covered, for 15 minutes. Add remaining ingredients and cook about 10 minutes longer, until potatoes are tender. Serves 4.

Fried Parsley and Parsley Butter

Parsley, a member of the carrot family and an excellent source of vitamins A and C, comes in a number of varieties, identified primarily by the shape of the leaves. The best-known are curly parsley and what is called common or Italian parsley, which has a flat leaf. Most cooks use parsley primarily as a garnish or seasoning for cooked dishes, but it is also an excellent primary ingredient for breads, jellies, sauces, and butters, and can also be fried by itself. Here are two inviting parsley recipes.

Fried Parsley
Wash parsley and dry very well to remove all moisture. Drop into hot deep fat or oil (375° F. on a frying thermometer) and fry until parsley is crisp and rises to the surface, probably less than a minute. Drain on absorbent paper and serve at once.

Parsley Butter

Use over cooked meats, fish, or vegetables, or with hot breads.

½ cup butter	Salt and freshly ground
1 tablespoon fresh lemon juice	pepper to taste
⅓ cup chopped fresh parsley	

Cream butter in a small bowl. Beat in lemon juice a little at a time. Add parsley and salt and pepper. Chill until ready to use. Makes about ¾ cup.

GREENS & LEAFY VEGETABLES

Dried Vegetables

Dried vegetables have nourished mankind for centuries. The broad—or fava—bean and the lentil were known in the Mediterranean area in prehistoric times, and the soybean is a very ancient Oriental food. The Indians of America cultivated a diverse number of beans, which grew in exceptional colors, sizes, and shapes, and were dried to last throughout the year.

Dried vegetables come in fascinating variety: black or turtle beans, black-eyed or cow peas, chick-peas, lentils, split peas, and beans—small white pea or navy, yellow-eye, cranberry, red and white marrow, pinto, red and white kidney, giant and small lima, Great Northern, and soy.

It is important to realize that dried vegetables are concentrated sources of energy with a high degree of protein that can match that of meat. They also contain a number of minerals, worthwhile amounts of iron, calcium and phosphorus, and are rich in the B vitamins. They are also widely available at reasonable prices.

Dried vegetables, also called legumes, are very versatile. Their cookery has been highly developed over the years, and there are truly treasured soups, stews, casseroles, and other dishes made with them. Early American settlers relied on a variety of dried beans for staple fare. Bean dishes were particularly important during the long winter months, since they supplied essential protein and a good rib-sticking quality. Our land would not have been settled without these foods.

All dried vegetables must be soaked in water before they are cooked to restore the liquid lost during drying. (Exceptions are split peas, lentils, and beans marked "quick-cooking" which need no presoaking). There are two methods of soaking. One is the old-fashioned overnight method; the other is the "quick" method, which is equally effective. In this second method the vegetables are covered with cold water, brought to a boil, allowed to cook for 2 minutes, then removed from the heat and left to stand in the same water for an hour. The vegetables are then simmered until mealy-tender. The soaking water should be used for the cooking.

Dried vegetables should be simmered and never boiled, as fast cooking causes them to burst. To keep down foam when cooking, add 1 tablespoon vegetable oil or butter for each cup of peas or beans.

The following recipes are especially good for family and informal company meals, and cover a wide selection ranging from appetizers to accompaniments.

Esau's Lentil Pottage

This is one version of the famous soup-stew in the Bible, a most ancient dish but one that is still inviting and nourishing.

1 cup dried lentils	2 garlic cloves, crushed
2 quarts water	3 tablespoons olive or vegetable oil
Salt and pepper to taste	
1 medium-sized onion, peeled and diced	3 tablespoons vinegar
	1/2 teaspoon cayenne pepper

Wash and pick over lentils. Combine with water, salt and pepper, in a large kettle. Bring to a boil. Lower heat and cook slowly, covered, about 1 1/2 hours, or until lentils are tender. Meanwhile, sauté onion and garlic in heated oil in a skillet until tender. When lentils are tender, add sautéed onion mixture and vinegar and cayenne, and cook 5 minutes longer. Correct seasoning, if necessary. Serves 6.

Cuban Black Bean Soup

Caribbean cooks are partial to dried black beans which they prepare in many interesting dishes, often in combination with rice. A most treasured specialty is this nourishing soup.

- 1 pound dried black beans, washed and picked over
- 2 quarts water
- 2 medium-sized onions, peeled and chopped
- 2 or 3 garlic cloves, crushed
- 1 medium-sized green pepper, cleaned and diced
- ½ cup olive or vegetable oil
- 2 medium-sized bay leaves
- Salt and pepper to taste
- 2 tablespoons wine vinegar
- About 3 cups hot boiled rice
- About 1 cup chopped onion (optional)

Cover beans with water in a large kettle. Bring to a boil; boil 2 minutes. Let stand, covered, 1 hour. Sauté onions, garlic, and green pepper in heated oil in a skillet until tender. Add to beans with bay leaves and salt and pepper. Bring to a boil and simmer, covered, 2 hours, or until beans are tender. Mix in vinegar and cook 5 minutes longer. Remove and discard bay leaves. Serve in soup bowls spooned over hot boiled rice and topped, if desired, with chopped raw onion. Serves 8 to 10.

French Canadian Split-Pea Soup

Also called habitant pea soup, this is a staple winter supper dish in eastern Canada and northern Maine. Serve with hot corn bread.

- 2 cups dried yellow peas, washed and picked over
- 2 quarts water
- ⅓ to ½ pound salt pork, blanched and diced
- 1 medium-sized onion, peeled and diced
- 1 medium-sized carrot, scraped and diced
- 1 tablespoon minced fresh parsley or ½ teaspoon dried sage
- Salt and pepper to taste

Soak peas in water in a large kettle overnight or for about 9 hours. Add remaining ingredients and bring to a boil. Lower heat and simmer, covered, until tender, about 1½ hours. Serves 6 to 8.

Near Eastern Chick-Pea Purée

This ancient purée, called *hummus,* can be served as a dip with bread or as an accompaniment to fish or shish kebab. Chick-peas marry well with sesame paste to make an unusual and desirable flavor.

- 1 can (20 ounces) chick-peas
- 2 garlic cloves, crushed
- ½ cup sesame paste (tahini) *
- Juice of 1 large lemon
- ½ cup water
- Salt to taste
- 2 tablespoons olive oil
- 2 tablespoons chopped fresh parsley
- 1 teaspoon paprika

Drain and rinse chick-peas. Mash in a bowl or whirl in a blender. Add garlic, sesame paste, lemon juice, water, and salt. Mix well. Chill. Turn onto a plate. With a finger, make a well in the center and pour oil into it. Sprinkle top with parsley and paprika. Serve as a dip with pieces of Arabic bread or crackers to scoop up the purée. Serves 6 to 8.

* *Tahini* can be purchased in Middle Eastern or specialty food stores.

Louisiana Red Beans

The red bean is particularly popular in the South and Southwest. Monday in Louisiana is red-bean and rice day, and this classic dish appears on that day on all restaurant menus in New Orleans.

1	pound dried red beans	2	tablespoons lard or shortening
6	cups water	½	teaspoon dried thyme
1	ham bone or ham hock (optional)		Salt and pepper to taste
			About 2 cups hot boiled rice
2	medium-sized onions, peeled and minced		Garnish: chopped scallions (optional)
2	garlic cloves, crushed		

Wash and pick over beans. Put beans and water in a kettle. Bring to a boil. Boil 2 minutes. Remove from heat and leave, covered, 1 hour. Add ham bone and return to heat. Cook slowly, covered, for 1 hour. Meanwhile sauté onions and garlic in lard in a skillet until tender. Add to beans with thyme, salt and pepper. Continue cooking slowly until beans are tender, about 25 minutes. Serve ladled over hot boiled rice and garnish, if desired, with chopped scallions. Serves 6.

Fool from Egypt

Fool, dried white beans, are customarily prepared in this manner in Egypt although sometimes they are cooked with lentils and mashed to a purée. The dish is served as an accompaniment to lamb or other meat.

2	cups dried white beans, washed and picked over	3	tablespoons fresh lemon juice
3	garlic cloves, crushed	1	teaspoon cayenne
½	cup chopped onions		Salt and pepper to taste
½	cup olive oil	½	cup chopped fresh coriander or parsley

Put beans in a large kettle and cover with water. Bring to a boil and cook, covered, for 2 minutes. Remove from heat and let stand, covered, 1 hour. Bring again to a boil. Lower heat and cook slowly, covered, until beans are tender, about 1½ hours. Add more water while cooking, if needed. Drain and turn into a large serving bowl. Add remaining ingredients and mix well. Chill. Serves 8.

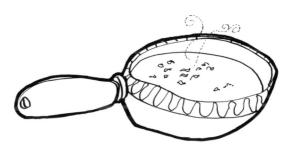

Indian Lentil-Rice Pot

In India the family cook selects a particular kind of lentil, which grow in dozens of varieties, and rice, of which there are many kinds, to prepare the family's dishes, and often the two are combined to make a filling, nourishing specialty such as this one.

1 medium-sized onion, peeled and chopped	½ teaspoon ground cinnamon
1 garlic clove, crushed	1 teaspoon ground cardamom
2 tablespoons peanut or vegetable oil	Salt and pepper to taste
	About 4 cups water
1 teaspoon turmeric powder	1 cup uncooked long-grain rice
¼ teaspoon ground cloves	1 cup dried lentils
	1 cup plain yogurt at room temperature

Sauté onion and garlic in heated oil in a large saucepan until tender. Stir in spices and salt and cook 1 minute. Add 4 cups water and bring to a boil. Mix in rice and lentils and lower heat. Cook slowly, covered, about 30 minutes, or until rice and lentils are tender and liquid is absorbed. Check to see if more water is needed during cooking. Stir in yogurt and remove from heat. Serve garnished with sliced hard-cooked eggs and crisply fried onion rings, if desired. Serves 8 to 10.

Turkish Taverna Beans

Staple fare in Turkish informal restaurants, or tavernas, is a dish of well-flavored cold white beans served with roast, skewered, or grilled meat and fish. These beans are also good for an outdoor meal.

2 cups dried white beans, washed and picked over	1 cup olive or vegetable oil
4 garlic cloves, crushed	1 teaspoon salt
2 medium-sized onions, peeled and chopped	1 tablespoon wine vinegar
	1 teaspoon sugar
2 medium-sized carrots, scraped and diced	3 tablespoons chopped fresh parsley
2 medium-sized stalks celery, cleaned and chopped	

Cover beans with water; bring to a boil; boil 2 minutes. Remove from heat. Let stand, covered, 1 hour. Simmer, covered, for 1 hour, adding more water, if needed. Add garlic, onions, carrots, celery, oil, and salt. Continue to cook slowly until beans are just tender, about 30 minutes longer. Stir in vinegar and sugar. Cook another 5 minutes. Cool. Serve cold in a large bowl garnished with parsley. Serves 8 to 10.

Boston Baked Beans

Boston has long been famous for its baked beans, which were once prepared for Sabbath meals but became traditional fare in New England with brown bread for Saturday night supper. This is the classic recipe, but it is subject to a number of variations.

2 cups dried pea beans, washed and picked over	1 medium-sized onion, peeled
	½ cup dark molasses
6 cups water	½ teaspoon dry mustard
¾ pound salt pork	1 teaspoon salt

116   *The Delectable Vegetable*

Put beans and water in a kettle. Boil 2 minutes. Let stand, covered, 1 hour. Then cook very slowly, covered, until beans are just tender, 1 hour or longer. Drain, reserving liquid. Put beans in a bean pot. Cut through pork rind every $1/2$ inch, penetrating about 1 inch deep. With a spoon, make a hollow in beans and put onion and pork in it, leaving top of pork exposed. Combine remaining ingredients with reserved bean liquid; pour over beans. Add enough water to cover ingredients. Bake, covered, at 300° F. for about 8 hours, or until beans are tender. Replenish water during cooking, if necessary, as beans throughout the cooking should be covered with liquid. Serves 6 to 8.

Italian White Beans with Tuna Fish

A good bean dish for a summer meal.

4 cups canned cannellini or other cooked or canned white beans, drained	$1/2$ cup chopped fresh parsley
	2 tablespoons fresh lemon juice
	Salt and pepper to taste
$1/2$ cup chopped onions	1 can (7 ounces) tuna fish in oil

Rinse canned beans quickly in cold water. Combine with onions, parsley, lemon juice, salt and pepper. Spoon onto a plate or into a shallow bowl. Divide tuna into chunks and arrange over bean mixture. Sprinkle with oil from tuna. Chill. Serves 6 to 8.

Russian Kidney Beans

The Georgians of southern Russia are devoted to beans, no matter what their color—white, green, or red. They often serve red kidney beans with a pounded walnut and red-pepper sauce or with this simple marinade.

2 cups cooked or canned kidney beans, drained	2 tablespoons red wine vinegar
½ cup chopped scallions, with tops	3 tablespoons chopped fresh dill or parsley
½ cup olive or vegetable oil	Salt and pepper to taste

Combine all ingredients and leave to marinate 2 hours or longer. Serves 4 to 6.

Southwestern Frijoles

Frijoles, as beans are commonly called in our Southwest, are generally pinto beans cooked with seasonings and salt pork or bacon.

1 pound pinto beans, washed and picked over	½ teaspoon dried oregano
6 cups water	About ½ pound salt pork
1 garlic clove	Salt to taste

Put beans and water in a kettle and soak overnight, or for about 8 hours. Add remaining ingredients and cook slowly, covered, until tender, about 1½ hours. Serves 6 or 8.

Note: Hot red chilis or chili powder can be added to the beans, if desired.

Refried Beans

Frijoles refritos, refried beans, are considered in our Southwest the best of all possible bean dishes.

Heat ⅓ cup bacon drippings or lard in a heavy pot or skillet. Add 4 cups cooked pinto beans, a few at a time. Mash well as beans are added. Moisten with a little bean liquid to make beans mushy. Cook over fairly high heat until crisp on bottom and fat is absorbed. Serve sprinkled with grated or cubed Monterey Jack or Cheddar cheese and/or chopped raw onions. Serves 4.

Dutch Bean Hutspot

The Dutch prepare many excellent hearty stews that are generally a combination of vegetables and inexpensive cuts of meat, and include often dried beans. This is a simplified version of Hutspot that could well be served for supper.

About ¼ pound salt pork
4 medium-sized onions, peeled and halved
8 medium-sized carrots, scraped and sliced
8 medium-sized potatoes, peeled and halved
Salt and pepper to taste
4 cups cooked or canned white beans
¼ cup chopped fresh parsley

Put salt pork, onions, carrots, potatoes, and salt and pepper in a large saucepan. Cover with water and bring to a boil. Lower heat and cook slowly, covered, 35 minutes. Add beans and parsley and cook another 10 minutes. Serves 4 to 6.

Lima Beans in Tomato Sauce

An easy-to-prepare dish that can be served with barbecued or fried chicken, grilled hamburgers or frankfurters.

1 medium-sized onion, peeled and chopped	1 can (1 pound) tomatoes, undrained
2 tablespoons vegetable oil	Salt and pepper to taste
1 teaspoon chili powder	4 cups cooked lima beans
¼ cup brown sugar	1 tablespoon cider vinegar

Sauté onion in heated oil until tender. Add chili powder, brown sugar, tomatoes, salt and pepper. Break up tomatoes with back of a spoon. Cook slowly, uncovered, for 10 minutes. Stir in beans and vinegar and cook another 5 minutes. Serves 4.

Greek Baked Chick-Peas

An innovative way of preparing the nutty-flavored chick-pea.

1 large onion, peeled and chopped	¼ cup water
2 garlic cloves, minced	½ teaspoon dried oregano
1 tablespoon olive or vegetable oil	3 tablespoons chopped fresh parsley
2 large tomatoes, peeled and chopped	Salt and pepper to taste
2 tablespoons tomato paste	2 cups cooked or canned chick-peas, drained

Sauté onion and garlic in oil in a skillet until tender. Mix in tomatoes, tomato paste, water, oregano, parsley, salt and pepper. Cook slowly, covered, 5 minutes. Put chick-peas in a shallow baking dish. Pour tomato sauce over them. Bake, covered, in a preheated 350° F. oven for 30 minutes. Serves 4.

Picnic Herbed White Beans

4 cups cooked white beans, drained and cooled	1 teaspoon dried thyme
⅓ cup olive or vegetable oil	Salt and pepper to taste
3 tablespoons vinegar	2 large tomatoes, peeled and cut into wedges
¾ cup chopped scallions, with tops	2 green peppers, cleaned and cut into strips
½ cup chopped fresh parsley	

Combine all ingredients, except tomatoes and green peppers, in a large bowl and mix well. Chill. Serve garnished with tomatoes and peppers. Serves 6.

Yugoslavian Cold White Beans

The Macedonians of Yugoslavia are famous for their piquant, well-flavored white-bean dishes served hot or cold at almost all family and company meals.

2 cups cooked white beans, drained and cooled	2 medium-sized green peppers, peeled and chopped
½ teaspoon cayenne pepper or paprika	1 teaspoon minced hot red peppers
Salt and pepper to taste	2 tablespoons minced fresh parsley
1 large red onion, peeled and cut into rings	Juice of 1 lemon
2 medium-sized tomatoes, peeled and chopped	About ⅓ cup olive or vegetable oil

Spoon beans into a bowl or onto a plate. Season with cayenne, salt and pepper. Top with onion rings, tomatoes, peppers, and parsley. Sprinkle with lemon juice and oil. Serves 4.

Mexican Chick-Peas

1 large onion, peeled and chopped	1 can (1 pound) tomatoes, undrained
2 garlic cloves, crushed	¼ cup tomato paste
2 tablespoons olive or vegetable oil	Salt and pepper to taste
1 tablespoon chili powder	2 cups cooked or canned chick-peas, drained
½ teaspoon dried oregano	¼ cup chopped fresh parsley
1 medium-sized bay leaf	

Sauté onion and garlic in heated oil in a saucepan until tender. Add chili powder, oregano, bay leaf, tomatoes, tomato paste, salt and pepper. Cook, uncovered, 15 minutes. Add chick-peas and parsley, and cook another 5 minutes. Serves 4.

Dal from India

Dal, a thick or thin purée, is made in India with many kinds of lentils and is a nourishing family staple.

½ pound dried lentils, washed	2 tablespoons chili powder
1 medium-sized onion, peeled and chopped	2 teaspoons coriander powder
4 cloves garlic	Salt and pepper to taste
2 tablespoons turmeric powder	½ cup tomato paste
	3 cups chicken broth or water

Combine ingredients in a saucepan and bring to a boil. Lower heat and simmer until lentils are tender, about 35 minutes. Purée or whirl in a blender, thinning with a little water, if desired. Serves 4 to 6.

Southern Hoppin' John

This mixture of black-eyed peas and rice is a traditional Southern New Year's Day dish, which is eaten to ensure good luck during the coming year.

1 cup dried black-eyed peas	1 cup uncooked rice
6 cups water	Salt and pepper to taste
1 medium-sized onion, peeled and chopped	½ to 1 teaspoon cayenne pepper
¼ pound piece bacon or salt pork	

Put peas and water in a kettle and bring to a boil; boil 2 minutes. Remove from heat and let stand, covered, 1 hour. Add onion, bacon or pork, and rice, salt, pepper, and cayenne. Bring again to a boil and then simmer, covered, about 1 hour, until peas and rice are tender. Put rice and peas into a warm bowl, and serve garnished with the bacon or pork, cut into slices. Serves 6.

Curried Lentils from Pakistan

Potatoes, eggplant, peas, or cauliflower can be added to this recipe.

2 medium-sized onions, peeled and finely chopped	2 teaspoons chili powder
	Salt and pepper to taste
2 garlic cloves, crushed	½ pound dried lentils
3 tablespoons vegetable oil	3 cups chicken broth or water
3 tablespoons curry powder	1 cup plain yogurt at room temperature
1 tablespoon turmeric powder	

Sauté onions and garlic in heated oil in a saucepan until tender. Add curry, turmeric, and chili powders, salt and pepper. Cook 1 minute. Add lentils and chicken broth or water and cook slowly, covered, until tender, about 40 minutes. Stir in yogurt and leave over low heat long enough to get hot. Serves 4 to 6.

Brazilian Black Beans

Frijoles negros, black beans, are staple fare in Brazil and are prepared in a festival dish with several kinds of meat and sometimes greens, such as collards or kale. This is a simplified version of the traditional dish.

- 1 pound dried black beans, washed and picked over
- 6 cups water
- 1 piece salt pork or bacon, about 1/3 pound
- 2 bay leaves
- 2 teaspoons salt
- 2 tablespoons olive or vegetable oil
- 1 medium-sized onion, peeled and chopped
- 2 garlic cloves, crushed
- 2 tomatoes, peeled and chopped
- 3 tablespoons chopped fresh parsley

Put beans and water in a kettle. Bring to a boil; boil 2 minutes. Remove from heat and let stand, covered, for 1 hour. Add pork or bacon, bay leaves, and salt to beans and bring to a boil. Lower heat and cook slowly, covered, 1 hour. Heat oil in a skillet. Add onion, garlic, and tomatoes and sauté 1 minute. Add to beans and continue to cook until tender, about 40 minutes longer. Stir in parsley. Serve, if desired, over hot cooked rice. Serves 6 to 8.

American Cassoulet

A modern American version of the famous French white bean, meat, and poultry dish, baked traditionally for hours in an earthenware pot. A good main course for a winter dinner.

- 1 pound dried Great Northern beans, washed and picked over
- 6 cups water
- 1 cup chopped celery
- 1 cup chopped scraped carrots
- 3 chicken bouillon cubes
- 2 teaspoons salt
- 1 cut-up frying chicken, about 3 pounds
- Pepper
- 3 tablespoons vegetable oil
- 1 smoked sausage, about 1 pound, cut into 1-inch slices
- 1 large onion, peeled and chopped
- 1 or 2 garlic cloves, crushed
- 1 can (15 ounces) tomato sauce
- 1/2 teaspoon dried oregano
- 1/3 cup chopped fresh parsley

Put beans and water in a large kettle. Bring to a boil; boil 2 minutes. Remove from heat and let stand, covered, 1 hour. Add celery, carrots, bouillon cubes, and salt. Bring to a boil again. Lower heat and cook slowly, covered, until beans are tender, about 1 hour. Meanwhile, wipe dry chicken pieces and sprinkle with salt and pepper. Fry in heated oil in a skillet until golden on all sides. Remove pieces to a heavy casserole. Add sausage slices and fry. Remove to casserole. Add onion and garlic to the drippings, and sauté until tender. Mix in tomato sauce and oregano. Season with salt and pepper. Cook slowly, uncovered, 10 minutes. When beans are tender, drain, reserving liquid. Spoon beans and carrots and celery over the chicken and sausage. Add 1 cup bean liquid and tomato sauce. Bake, covered, in preheated 325° F. oven 1 hour, or until ingredients are tender and casserole is bubbly. Sprinkle with parsley. Serves 6.

Soyburgers

A nutritious substitute for the well-known meat burger.

2 cups cooked dried soybeans	1 tablespoon Worcestershire sauce
¼ cup minced onions	½ teaspoon salt
1 cup cooked brown or white rice	About 2 tablespoons vegetable oil
1 cup soft whole wheat or white bread crumbs	1 can (8 ounces) tomato sauce
2 eggs, beaten	

Mash or purée soybeans and combine with remaining ingredients, except oil and sauce, in a large bowl. Shape mixture into patties. Heat oil in a skillet and brown patties on both sides. Add tomato sauce and simmer, covered, for 20 minutes, or until tender. Serves 6 to 8.

Vegetables as Accompaniments

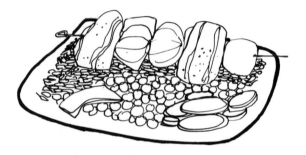

The traditional role for vegetables on our tables has been as an accompaniment for meats, poultry, or seafood. Too often the vegetable is cooked and served casually, and attracts little or no attention.

Vegetables served as accompaniments should be given a great deal of consideration and be carefully prepared as they can add greatly to the pleasure of a meal.

When pondering the choice of vegetable and whether to serve a single one or a dish made with several, consider color, texture, and flavor and what other foods you will serve with it. While some vegetables, such as green peas, seem to be agreeable companions for almost any food, this is by no means true of all. It is desirable to serve simple or bland vegetable dishes with intricate main courses, and creamy or spicy vegetables with a roast or with chops.

In each of the following vegetable recipes an appropriate main dish which it might accompany has been suggested, but they are intended as merely a guide.

Norwegian Spiced Red Cabbage

There are several varieties of reddish-purple cabbage that are usually prepared in northern European countries with a sweet-sour flavoring. It is advisable to include something acid, such as vinegar or lemon juice, when cooking red cabbage, so that the vegetable does not turn an unattractive purple. Red cabbage requires longer cooking than green cabbage. This is a good accompaniment to pork and game.

- 1 medium-sized head red cabbage, about 2 pounds
- ¼ cup butter or margarine
- 1 large tart apple, peeled, cored, and sliced
- 1 tablespoon brown sugar
- 1 teaspoon caraway seeds
- Salt to taste
- 3 tablespoons red wine vinegar
- ¼ cup red-currant jelly or juice

Remove any wilted outer leaves from cabbage and shred, discarding stalk and tough ribs. Melt butter in a saucepan and sauté cabbage in it. Add apple, sugar, caraway seeds, salt, and vinegar. Cook slowly, covered, for 45 minutes. Add jelly, or juice, and cook 10 minutes longer. Serves 6.

Stir-Fry Snow Peas with Mushrooms and Bamboo Shoots

An Oriental dish to accompany poultry or beef.

- 2 tablespoons peanut or vegetable oil
- 1 large onion, peeled and cut into shreds
- 1 thin slice ginger root (optional)
- 2 cups sliced fresh mushrooms
- ½ cup sliced bamboo shoots
- 2 packages (6 ounces each) frozen snow peas or pea pods, defrosted
- 2 tablespoons chicken broth

Heat oil in a wok or skillet. Add onion, and ginger root if desired, and sauté until tender. Add mushrooms and bamboo shoots and stir-fry 2 minutes. Add snow peas and broth and cook, stirring, over fairly high heat about 2 minutes. Serve at once. Serves 4.

Flemish Brussels Sprouts with Cheese

Brussels sprouts, members of the cabbage family that look like miniature cabbages, are believed to have originated in Brussels, the capital of Belgium, in the late eighteenth century. The tiny vegetables grow in considerable quantity on a thick stem and are thus called sprouts. Excellent flavorings for them are nutmeg and cheese. Serve with roast beef or veal.

- 2 pints fresh or 2 packages (10 ounces each) frozen Brussels sprouts
- Boiling salted water
- 3 tablespoons melted butter or margarine
- 1 cup grated Swiss, Parmesan, or Edam cheese
- 2 teaspoons fresh lemon juice
- $1/8$ teaspoon grated nutmeg
- Pepper to taste

If fresh sprouts are used, wash, remove wilted leaves, and trim stems. Cook fresh or frozen sprouts in a little boiling salted water until just tender—about 6 minutes for fresh sprouts. Drain. Arrange in a shallow baking dish. Combine remaining ingredients and pour over sprouts. Bake, covered, in a preheated 350° F. oven for 20 minutes. Serves 4.

Fried Green Tomatoes with Milk Gravy

An old-fashioned country dish, a good accompaniment to fried chicken.

- 6 medium-sized green tomatoes
- About $1/2$ cup all-purpose flour
- Salt and pepper to taste
- About 5 tablespoons bacon drippings or shortening
- 1 cup milk

Wash tomatoes; remove stem ends; cut into thick slices. Dredge in flour, seasoned with salt and pepper. Heat 2 tablespoons bacon drippings in a skillet and quickly sauté tomato slices, a few at a time, until golden brown on both sides. Remove to a hot platter when cooked and keep warm, continuing to sautée and adding fat as needed, until all the tomatoes are cooked. Mix 2 tablespoons flour into drippings. Add milk and cook slowly, stirring constantly, until thick and smooth. Spoon over tomatoes. Serves 4 to 6.

Indian Green Beans with Coconut

Green beans have been and still are called snap, pole, and bush beans, as well as French and string beans, the last name having come into use because until quite recently green beans did have strings, which had to be removed by hand before cooking. Now most green beans are of the stringless variety. Serve with poultry or lamb.

2 packages (10 ounces each) frozen cut-up green beans
3 tablespoons butter or margarine
2 tablespoons ground turmeric
Salt and pepper to taste
1/4 cup grated coconut, preferably unsweetened

Cook beans according to package directions until just tender. Drain. Meanwhile, melt butter in a saucepan. Mix in turmeric, salt and pepper. Cook 1 minute. Add cooked beans and coconut. Mix well and serve. Serves 6 to 8.

Old-Fashioned Turnip Custard

Serve with roast chicken or turkey.

2 eggs	1 teaspoon grated onion
2 cups mashed cooked turnips, white or yellow	2 tablespoons melted butter or margarine
1 cup milk, scalded	Salt and pepper to taste

Beat eggs with a fork in a large bowl. Stir in remaining ingredients and spoon into a round greased baking dish, about 1 quart in size in a pan of hot water. Bake in preheated 350° F. oven for 1 hour, or until a tester inserted into center comes out clean. Serves 6.

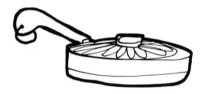

Flemish Carrots

Carottes flamandes, as it is called in its native land, is an appealing, delicately flavored carrot dish that is especially good with chicken or veal.

3 tablespoons butter or margarine	2 tablespoons water
8 medium-sized carrots, scraped and thickly sliced	½ cup heavy cream
	¼ teaspoon freshly grated nutmeg
2 teaspoons sugar	2 tablespoons chopped fresh parsley
Salt and pepper to taste	

Melt butter in a saucepan. Add carrots, sugar, salt, pepper and water. Cook slowly over moderate heat, covered, until carrots are tender, about 15 minutes. Add remaining ingredients and bring to a quick boil. Remove at once from heat. Serves 4.

Lima Beans, Texas Style

Serve at an outdoor barbecue with pork, steaks, or hamburgers.

- 1 medium-sized onion, peeled and chopped
- 1 garlic clove, crushed
- 1 tablespoon vegetable oil
- 1 tablespoon chili powder
- 1 can (8 ounces) tomato sauce
- 2 teaspoons sharp mustard
- 2 teaspoons Worcestershire sauce
- 1/2 teaspoon dried oregano
- Salt and pepper to taste
- 2 cups fresh or frozen baby lima beans

Sauté onion and garlic in oil in a medium saucepan until tender. Mix in chili powder and cook 1 minute. Add remaining ingredients, except beans, and cook slowly, uncovered, 5 minutes. Add beans and cook until tender, about 12 minutes. Serves 4.

Risi e Bisi from Venice

Rice and peas are served together in Venice either as a soup, with considerable liquid, or as in this recipe, as an accompaniment to meat, fish, and poultry with practically no liquid.

- 1/3 cup chopped cooked ham
- 1 small onion, peeled and minced
- 4 tablespoons butter or margarine
- 1/2 cup raw long-grain rice
- 1 cup chicken broth
- Salt and pepper to taste
- 2 cups fresh or frozen green peas
- About 1/3 cup grated Parmesan cheese

Combine ham, onion, and 2 tablespoons heated butter in a medium-sized saucepan and sauté until onion is tender. Add rice and sauté until grains are translucent, about 5 minutes. Add broth, salt and pepper, and bring to a boil. Lower heat and cook slowly, covered, about 20 minutes, or until rice grains are just tender and liquid has been absorbed. Add peas during last 10 minutes of cooking if fresh, and last 5 minutes if frozen. Serve with grated cheese. Serves 4.

Turkish Cold Green Beans

An excellent dish for entertaining as it can be prepared beforehand. Serve with any kind of meat or poultry.

- 2 packages (10 ounces each) frozen green beans, French cut
- 1 large onion, peeled and chopped
- 2 large tomatoes, peeled and chopped
- ½ cup olive or vegetable oil
- 1 tablespoon sugar
- Salt and pepper to taste
- 3 tablespoons chopped fresh parsley

Combine all ingredients, except parsley, in a saucepan. Cook slowly, covered, for 25 minutes. Remove from stove and cool. Serve cold sprinkled with parsley. Serves 6 to 8.

Scotch Braised Leeks

The Scots, as well as the Welsh and Irish, are devoted to leeks, which they prepare in interesting variations and serve with lamb, particularly roast leg of lamb.

- 12 leeks
- 1 medium-sized onion, peeled and minced
- 3 tablespoons butter or margarine
- 1¼ cups beef consommé
- Dash of grated nutmeg
- Salt and pepper to taste
- Toasted English Muffins (optional)
- About ⅓ cup grated Parmesan cheese (optional)

Cut green tops from leeks. Trim roots. Cut white parts in half and wash well to remove sand. Sauté onion in heated butter in a skillet until tender. Add leeks and remaining ingredients. Cook slowly, covered, until tender, about 15 minutes. Serve plain or on toasted English muffins sprinkled with grated Parmesan cheese. Serves 4.

Moroccan Artichokes with Lemon-Olive Dressing

A flavorful dish to serve with roast beef, veal, or lamb.

- 2 packages (9 ounces each) frozen artichoke hearts
- ½ cup olive or vegetable oil
- Juice of 2 lemons
- ½ cup chopped green onions, with tops
- ½ cup sliced pitted black olives
- Salt and pepper to taste
- 6 large lettuce leaves

Cook artichoke hearts according to package directions. Drain and chill. Combine remaining ingredients, except lettuce, and chill. When ready to serve, arrange artichoke hearts on lettuce leaves on small individual plates. Spoon dressing over artichokes. Serves 6.

Australian Orange Carrots

A good accompaniment to roast lamb or lamb chops.

- 4 medium-sized carrots, washed and scraped
- Boiling salted water
- 1 tablespoon sugar
- 1 teaspoon cornstarch
- ¼ teaspoon ground ginger
- ¼ cup orange juice
- 2 tablespoons butter or margarine
- Salt to taste

Slice carrots crosswise, about 1 inch thick. Cook in boiling salted water until just tender, about 20 minutes. Drain. Meanwhile, combine sugar, cornstarch, and ginger in a small saucepan. Add orange juice and cook slowly, stirring constantly, until thick and smooth. Stir in butter and season with salt. Pour over hot carrots and toss to coat evenly. Serves 4.

Alsatian Sauerkraut

Serve with roast pork or baked ham.

- 1 pound sauerkraut
- 1 large onion, peeled and chopped
- 2 tablespoons bacon fat or shortening
- 1 tart medium-sized apple, peeled, cored, and chopped
- 3 whole peppercorns
- 5 juniper berries
- ¾ cup dry white wine or chicken broth

Soak sauerkraut in cold water for 15 minutes. Squeeze dry between hands to release all liquid. Sauté onion in fat in a saucepan until tender. Add sauerkraut and toss with a fork. Cook for 5 minutes, stirring occasionally. Add remaining ingredients and cook slowly, covered, for 40 minutes. Remove and discard peppercorns and juniper berries. Serves 4.

Vermont Acorn Squash

A favorite winter vegetable, the acorn squash, a native American food, takes its name from its shape, but it is also known as Table Queen and Des Moines (because it was first grown commercially in the area of that city). The yellow-orange flesh bakes well, whether kept whole or cut in halves or rings, and it is often seasoned with spices and such sweetenings as maple syrup. Serve with roast turkey or chicken.

Scrub well 2 acorn squash, about 1 pound each. Cut each in half crosswise. Remove seeds and fibers. Dot insides with butter and sprinkle with salt and pepper. Add a little hot water and bake in preheated 425° F. oven for 25 minutes. Remove from oven. Fill partially or brush insides with maple syrup, the amount being according to taste. Return to oven and cook until tender, about 25 minutes longer. Serve in shells. Serves 4.

Greek Vegetable Medley

Serve with broiled or baked fish.

- 2 medium-sized onions, peeled and sliced
- 2 garlic cloves, crushed
- 3 tablespoons olive or vegetable oil
- 4 medium-sized potatoes, peeled and cubed
- 3 zucchini, ends trimmed and sliced
- 2 cups cut-up green beans
- About 1 cup tomato juice
- 2 tablespoons fresh lemon juice
- 1 teaspoon dried oregano
- Salt and pepper to taste
- 1/4 cup chopped fresh parsley

Sauté onions and garlic in heated oil in a saucepan until tender. Add remaining ingredients, except parsley, and cook slowly, covered, about 25 minutes, or until vegetables are tender. Add a little more tomato juice during cooking, if needed. Stir in parsley. Serves 4.

Baked Tomatoes Provençale

Serve with any broiled meat, especially steak or chops.

- 6 firm ripe tomatoes of uniform size
- Salt and pepper to taste
- 1/4 cup minced shallots or scallions
- 2 or 3 garlic cloves, crushed
- 1/2 cup fine dry bread crumbs
- 1/2 teaspoon dried basil
- 1/2 cup chopped fresh parsley

Remove stems and cut tomatoes in halves crosswise. Carefully scoop out pulp and seeds and discard. Sprinkle tomato shells with salt and pepper and invert to drain. Combine next five ingredients and spoon into tomato halves. Place in a shallow greased baking dish. Bake in preheated 400° F. oven until golden and crusty on top, about 10 minutes. Serves 6.

Bohemian Creamed Kohlrabi

Another member of the cabbage family, the purplish-white kohlrabi, a name meaning cabbage turnip, is a great favorite in northern and central Europe. Its thick skin should be completely cut off before cooking. If the leaves are tender, they can also be cooked. Kohlrabi is a good accompaniment to pork dishes.

2 small kohlrabi
Boiling salted water
7 tablespoons butter or margarine
¼ cup light cream

Salt and pepper to taste
⅛ teaspoon grated nutmeg
¼ cup all-purpose flour
⅓ cup chopped fresh parsley

Cut leaves from kohlrabi and cook them in boiling salted water until tender, about 20 minutes. Drain and chop. Add 3 tablespoons butter and cream. Season with salt, pepper, and nutmeg. Keep warm. Meanwhile trim kohlrabi globular stems; peel and slice. Cook slices in 2 cups boiling salted water until tender, about 25 minutes. Drain, reserving liquid. Melt remaining butter in a saucepan and stir in flour. Season with salt and pepper. Add reserved vegetable liquid. Cook slowly, stirring, until thick and smooth. Add cooked kohlrabi slices and parsley; heat. Serve surrounded with cooked leaves. Serves 4

Connecticut Baked Jerusalem Artichokes

This native North American vegetable bears a confusing name because it has no visible resemblance to the globe artichoke nor has it anything to do with the city of Jerusalem. The name is derived from *Girasole*, the Italian word for sunflower, and it is a member of that family. Early Americans were very fond of this gnarled brown tuber, which has a white flesh with a sweetish flavor.

6 Jerusalem artichokes	2 cups milk
Boiling salted water	Salt and pepper to taste
3 tablespoons butter or margarine	½ cup buttered dry bread crumbs
3 tablespoons all-purpose flour	¼ cup grated Parmesan cheese

Wash and scrub artichokes. Do not peel. Cook, covered, in boiling salted water to cover about 20 minutes, or until tender. Drain, rub off skins, and slice. Arrange in a greased shallow baking dish. Melt butter in a saucepan and stir in flour. Cook 1 minute. Gradually add milk and cook slowly, stirring, until thick and smooth. Season with salt and pepper. Pour over artichokes. Sprinkle with bread crumbs and cheese. Bake in preheated 400° F. oven until hot and bubbly, about 12 minutes. Serves 4.

Swedish Mashed Rutabaga and Potatoes

Rutabaga, a root vegetable with yellow flesh, is also known as the Swedish turnip, yellow turnip, or swede, and is related to a number of vegetables. It can be cooked in any way that turnip is cooked and is a good winter vegetable.

1 medium-sized rutabaga	3 tablespoons light cream or milk
3 medium-sized potatoes, washed peeled, and quartered	2 teaspoons sugar
1 medium-sized onion, peeled and quartered	¼ teaspoon ground allspice
2 tablespoons butter or margarine	Salt and pepper to taste

Scrub rutabaga well and cut off outer skin. Cut into cubes. Cook, covered, in 1 inch of boiling water for 30 minutes. Add potatoes and onion and continue cooking until vegetables are tender, about 20 minutes. Drain and mash. Add remaining ingredients and mix well. Serve at once. Serves 4.

Bubble and Squeak from England

There are many versions of this curiously named dish. Today most of them are made with leftover cooked vegetables—cabbage, carrots, potatoes, or rutabagas—fried with or without meat until brown and crispy. According to the English, the name came about because of the cooking noises. There was first a bubble from the water, while the cabbage and potatoes boiled and then a squeak from the bacon fat, during the frying. Serve with pork or lamb.

- 2 slices thin bacon, chopped
- 1 large onion, peeled and chopped
- 2 cups cold cut-up cooked cabbage or Brussels sprouts
- 2 cups cold mashed potatoes

Fry bacon in a skillet. Add onion and sauté until tender. Combine cabbage or Brussels sprouts and potatoes and mash well. Turn into skillet and fry over a medium flame, pressing with a spatula, until golden brown. Turn over and cook on other side until golden. Serves 4.

South American Corn Pudding

Serve with roast beef or veal or pork dishes.

- 2 tablespoons butter or margarine
- 2 tablespoons minced onion
- 2 tablespoons all-purpose flour
- 2 tablespoons brown sugar
- 1 cup milk
- 2 cups canned whole kernel corn
- 1/2 teaspoon grated nutmeg
- Salt and pepper to taste
- 2 eggs, well beaten

Melt butter in a medium saucepan. Add onions and sauté until tender. Mix in flour and sugar. Cook 1 minute. Gradually add milk and then corn and seasonings. Cook slowly, stirring, for 2 minutes. Spoon some of hot mixture into eggs and blend well. Add to corn mixture. Spoon into buttered 1-quart baking dish. Set in a pan of water and bake in preheated 300° F. oven for about 1 hour, or until a knife inserted into mixture comes out clean. Serves 6.

Creole Wax Beans

Less common than the green bean, the wax bean derives its name from its shiny yellow color. A good summer dish to serve at a cookout with frankfurters or hamburgers.

2 slices thin bacon, chopped	¼ teaspoon dried basil
¼ cup minced onion	Salt and pepper to taste
¼ cup minced green pepper	4 cups cooked cut-up wax beans
¼ cup minced celery	2 tablespoons chopped fresh parsley
3 medium-sized tomatoes, peeled and chopped	

Cook bacon in a medium-sized saucepan until almost crisp. Add onion, green pepper, and celery. Sauté until tender. Mix in tomatoes, basil, salt and pepper, and cook slowly, uncovered, for 10 minutes. Add beans and parsley and leave on stove until hot. Serves 6.

New Hampshire Scalloped Potatoes and Parsnips

An early American favorite to serve with corned beef or meat loaf.

2 cups thinly sliced peeled potatoes	2 tablespoons all-purpose flour
1 medium-sized onion, peeled and chopped	Salt and pepper to taste
	2 tablespoons butter or margarine
2 cups thinly sliced peeled parsnips	About 1½ cups light cream or milk

Arrange potatoes, onion, and parsnips in layers in a greased 1½-quart baking dish, sprinkling each layer with flour, salt and pepper, and dotting with butter. Add enough cream or milk to almost cover ingredients. Bake, covered, in preheated 350° F. oven for 1 hour, or until vegetables are tender. Uncover to brown top during last 15 minutes of cooking. Serves 4.

Israeli Carrot-Sweet Potato Tzimmes

A good accompaniment for lamb or beef.

- 3 cups sliced cooked carrots
- 4 cups sliced cooked sweet potatoes
- 4 medium-sized tart apples, peeled, cored, and sliced
- ½ cup honey or brown sugar
- ½ cup hot water
- 2 tablespoons fresh lemon juice
- 2 teaspoons grated lemon rind
- Salt and pepper to taste

Arrange carrots, potatoes, and apples in a buttered baking dish, sprinkling each layer with a little honey or brown sugar, water, lemon juice and rind, salt and pepper. Bake, covered, in a preheated 350° F. oven until cooked, about 30 minutes. Uncover during last 10 minutes of cooking. Serves 6 to 8.

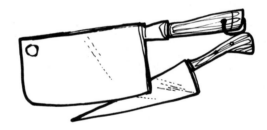

Salads

Of all the appealing vegetable dishes, salad, whether served as an appetizer or as a main or separate course, rates high on almost every American table. Little wonder, for salad is attractive, tasty, and nourishing, and can be prepared in many interesting ways.

Salad was probably first made in ancient times of edible wild herbs and greens seasoned only with salt. Indeed, our word salad is derived from *sal,* the Latin word for salt.

Eventually cooks began dressing raw and cooked vegetables with olive or sesame oils, vinegar or lemon juice, herbs, and other seasonings, and a fascinating array of salads emerged. Today there is an infinite variety of salad, ranging from simple to intricate combinations.

One of the most popular kinds in America is the tossed salad, generally made with such greens as lettuce, romaine, escarole, curly endive, or spinach, and sometimes other foods. It is usually served from a salad bowl that may or may not have been rubbed with garlic.

A tossed salad must be prepared with care and attention, or the result can be disastrous. Greens must be fresh, crisp, and tender, with any tough or bruised parts removed. After washing well, dry and refrigerate to maintain crispness until serving time. There should be no water clinging to them as it will dilute the dressing.

For salad dressing the cook has a wide choice among those that can be made at home and purchased. An important ingredient is the oil, which can be the more expensive olive or walnut oil or one of the less costly vegetable oils. Vinegar or lemon juice—or perhaps apple, pineapple, or one of the berry juices—used in the proportion of one part to three or four of oil, and herbs—basil, parsley, tarragon, or marjoram—are also important salad-dressing ingredients. Add the dressing to the greens just before tossing and serving.

Salads are excellent sources of nutritional value, carrying a good amount of vitamins and minerals in an appetizing form. If carefully selected, the ingredients of a salad can be low in calories.

Included in this section are a number of recipes for lesser-known but ever-appealing salads, displaying the versatility of the international cuisine.

Danish Vegetable Salad with Cheese

A good salad to serve with sandwiches for luncheon or supper.

1½ cups cold boiled potatoes, peeled and diced	About ⅓ cup mayonnaise
1½ cups cooked green peas	2 teaspoons sharp mustard
½ cup cubed Swiss or Gouda cheese	2 tablespoons cider vinegar
1 tablespoon minced gherkins or relish	Salt and pepper to taste

Combine potatoes, peas, cheese, and gherkins in a bowl. Blend together remaining ingredients. Add to vegetable mixture. Chill 1 hour or longer before serving. Serves 4.

Turkish Cauliflower Salad

Good to serve with lamb or grilled hamburgers.

1 package (10 ounces) frozen cauliflower	1/4 cup olive or vegetable oil
1/2 cup pitted halved black olives	2 tablespoons fresh lemon juice
1/2 cup chopped green peppers	1/4 teaspoon dried thyme
	Salt and pepper to taste

Cook cauliflower according to package directions until just tender. Cool. Cut into bite-size pieces. Combine with olives and peppers in a bowl. Mix together remaining ingredients and pour over vegetables. Chill. Serves 4.

Tomato Flower Salad

An excellent salad for a women's luncheon.

6 chilled medium-sized ripe tomatoes	2 hard-cooked eggs, peeled and chopped
6 large lettuce leaves	About 2 tablespoons mayonnaise
1/2 cup chopped celery	Salt and pepper to taste
1/2 cup chopped peeled cucumber	
1 tablespoon minced onion	
1 tablespoon chopped green pepper	

Cut stem ends from tomatoes. Cut each tomato four times from the top almost through to the bottom to form eight attached petals. Place lettuce leaves on six salad plates. Spread petals of each tomato and place on a lettuce leaf. Combine remaining ingredients, using enough mayonnaise to bind them, and spoon into tomato centers. Serves 6.

Norwegian Summer Salad

Serve with cold roast meat for a summer luncheon.

2 medium-sized boiled potatoes, peeled and sliced	2 tablespoons vinegar
	1/3 cup vegetable oil
2 hard-cooked eggs, peeled and chopped	2 teaspoons sharp mustard
	1 teaspoon sugar
1 small head leaf lettuce, cleaned and torn into bite-size pieces	Salt and pepper to taste
	2 medium-sized tomatoes, peeled and sliced

Mix potatoes, eggs, and lettuce in a salad bowl. Combine vinegar, oil, mustard, sugar, salt and pepper, and pour over vegetable-egg mixture. Toss lightly. Garnish with tomato slices. Serves 4.

Ozark Sweet-Sour Cole Slaw

A salad made of shredded raw cabbage, and sometimes other vegetables or foods, mixed with a piquant dressing has long been a staple dish in America. Now made in many versions, it was originally introduced to this country by Dutch colonists in New York.

2 tablespoons all-purpose flour	1 egg
1 teaspoon powdered dry mustard	Salt and pepper to taste
	1 cup milk
1/3 cup cider vinegar	3 cups shredded green cabbage
2 tablespoons sugar	1 cup grated carrots

Combine flour, mustard, vinegar, sugar, egg, and salt and pepper in a saucepan. Heat, stirring almost constantly, until thickened. Gradually add milk and continue cooking until smooth, about 2 minutes. Remove from heat and cool. Mix with cabbage and carrots. Refrigerate until ready to serve. Serves 6.

Oriental Bean-Sprout Salad

An interesting and attractive medley of vegetables with a soy-vinegar sauce to serve for a family meal.

- 2 cups canned bean sprouts, drained
- 1/2 cup thinly sliced white or red radishes
- 1 medium-sized cucumber, peeled and thinly sliced
- 6 scallions, with tops, sliced
- 2 medium-sized carrots, scraped and thinly sliced
- 1 tablespoon soy sauce
- 1 tablespoon vinegar
- 3 tablespoons peanut or vegetable oil
- 1/2 cup chopped blanched almonds
- Pepper to taste

Combine ingredients in a bowl. Serves 6.

Classic Greek Salad

A popular salad in Greece features romaine, black olives, and salty white Feta cheese in combination with as many other ingredients as the cook wishes to add. Here is the basic recipe.

- 1 large head romaine
- 1 medium-sized cucumber, peeled and sliced thinly
- 2 large tomatoes, peeled and cut into wedges
- 1 cup green pepper, cleaned and cut into strips
- 1/3 cup olive oil
- 2 tablespoons fresh lemon juice
- Salt and pepper to taste
- 1/2 cup crumbled Feta or farmers' cheese
- 12 pitted black olives

Wash, trim, and dry romaine. Break into small pieces. Combine with remaining ingredients and toss lightly. Serves 6.

Note: The amount of oil and lemon juice used may be a little more or a little less, according to taste.

Near Eastern Beet Salad

2 cups sliced cooked or canned beets, drained	1 teaspoon sugar
2 tablespoons wine vinegar	$1/3$ to $1/2$ cup plain yogurt
2 teaspoons prepared horseradish	Salt and pepper to taste

Combine ingredients and chill. Serves 4.

Viennese Wilted Cucumbers

Gurkensalat, Austria's most popular form of salad, is made in several ways. Garlic, oil, and/or paprika may be added to the following basic ingredients, if desired.

2 medium-sized cucumbers	White pepper to taste
Salt	2 teaspoons sugar
About $1/4$ cup white vinegar	

Peel cucumbers and cut off ends. Score lengthwise with a fork and cut into thin slices. Arrange in a bowl in layers, sprinkling each layer with salt. Refrigerate, covered, for 1 hour. Drain well. Combine remaining ingredients and pour over cucumbers. Refrigerate 1 hour before serving. Serves 6.

French Celeriac Rémoulade

The gnarled root vegetable called celeriac or celery root, is traditionally served in France as a salad for luncheon *hors d'oeuvre*. It is also a good accompaniment for pork dishes.

1 celery root, 1 to 1½ pounds	Salt and white pepper to taste
2 tablespoons olive or vegetable oil	About 1 cup mayonnaise
	1 tablespoon drained capers
2 teaspoons wine vinegar	1 garlic clove, crushed
⅛ teaspoon dry powdered mustard	¼ teaspoon dried tarragon
	¼ teaspoon dried chervil

Pare celery root, removing tops and roots, and cut into matchlike strips. Marinate in oil, vinegar, mustard, salt and pepper, for 2 hours. Drain and combine with remaining ingredients. Serves 4.

Peruvian Potato Salad with Cottage Cheese

An unusual salad featuring the white potato, which originated in Peru thousands of years ago and has been treasured there and elsewhere ever since.

6 medium-sized potatoes, well scrubbed	¼ cup milk
	1 tablespoon fresh lemon juice
2 tablespoons vegetable oil	Salt and pepper to taste
1 medium-sized onion, peeled and minced	Garnishes: Sliced radishes, chopped fresh parsley, halved pitted olives
2 to 3 tablespoons chili powder	
1 cup small-curd cottage cheese	

Peel, boil, and dice potatoes. Cool. Heat oil in a skillet and add onion. Sauté until tender. Add chili powder and cook 1 minute. Remove from heat and add cottage cheese, milk, lemon juice, salt and pepper. Beat until smooth. Arrange potatoes in a bowl or on a platter. Cover with cottage cheese sauce and decorate with garnishes. Serves 4 to 6.

Italian Tomato-Bread Salad

Serve with broiled or grilled meats, such as steak, shish kebab, or hamburgers.

8 tablespoons olive or vegetable oil	5 teaspoons wine vinegar
3 tablespoons butter or margarine	½ teaspoon dried basil
	Salt and pepper to taste
1½ cups 1-inch stale-bread cubes	2 tablespoons drained capers
4 large tomatoes, peeled and sliced	2 tablespoons chopped fresh parsley

Heat 3 tablespoons oil and the butter or margarine in a skillet. Add bread cubes and brown. Bake in preheated 275° F. oven for 25 minutes. Drain on absorbent paper. Arrange alternately with tomato slices in a deep dish. Mix vinegar, 5 tablespoons oil, basil, salt and pepper, and spoon over tomato-bread mixture. Sprinkle with capers and parsley. Serves 4.

Lithuanian Mixed-Vegetable Salad

1 cup sliced cooked carrots	2 tablespoons minced dill pickle
1 cup diced cooked beets	About 1 cup sour cream
2 cups sliced boiled potatoes	Salt and pepper to taste
1 cup cooked green peas	

Combine ingredients and chill. Serves 6 to 8.

German Carrot-Sauerkraut Salad

A good salad to serve with pork or game.

3 cups drained sauerkraut	2 tablespoons vinegar
1 cup thinly sliced carrots	½ teaspoon caraway seeds
1 medium-sized onion, peeled and chopped	2 teaspoons sugar
¼ cup vegetable oil	Salt and pepper to taste

Combine ingredients and mix well. Serves 4 to 6.

Salade Niçoise

An excellent summer main-course salad, particularly good for an outdoor meal.

2 large tomatoes, peeled and quartered	12 pitted black olives
3 cold boiled medium-sized potatoes, peeled and sliced	6 anchovy fillets, drained and chopped
1 cup cut-up cold cooked green beans	3 hard-cooked eggs, shelled and quartered
1 medium-sized Bermuda onion, peeled and sliced	⅓ cup olive or vegetable oil
	2 tablespoons wine vinegar
1 large green pepper, cleaned and cut into strips	1 teaspoon sharp mustard
	Salt and pepper to taste

Put first five ingredients in a large bowl. Arrange olives, anchovies, and eggs over vegetables. Combine remaining ingredients and mix well. Pour over salad and refrigerate 1 hour before serving. Mix again. Serves 4 to 6.

Tomato Salad Ring with Cucumbers

An attractive salad for a luncheon or buffet.

5 cups tomato juice	3 tablespoons plain gelatine
1 medium-sized stalk celery	1/3 cup cider vinegar
3 lemon slices	3 cups sliced, peeled cucumbers
2 small bay leaves	1/2 teaspoon dried chervil
Salt and pepper to taste	1 cup sour cream

Combine 4 cups tomato juice, celery, lemon slices, bay leaves, salt and pepper, in a saucepan and simmer, uncovered, 10 minutes. Strain. Stir gelatine into remaining 1 cup tomato juice. Add, with vinegar, to hot liquid. Stir to dissolve. Pour into a large ring mold. Chill until firm. Meanwhile, combine cucumbers, chervil, and sour cream. Chill. When ready to serve, unmold tomato ring onto a cold platter. Fill with cucumber mixture. Serves 8.

Italian Fennel Salad

An aromatic plant belonging to the carrot family, fennel is known in two major varieties. Common fennel is grown primarily for its leaves, and aromatic seeds, which are used as seasoning. Sweet fennel has a celery-like bulb with an anise flavor and is often called *finocchio*. The latter is excellent, either raw or cooked, and is particularly desirable as an ingredient for salad.

4 medium-sized bulbs sweet fennel	1 tablespoon fresh lemon juice
2 large tomatoes, peeled and sliced	3 tablespoons olive or vegetable oil
1 tablespoon drained capers	Salt and pepper to taste

Trim off green feathery tops and tough outer stalks of fennel. Cut into thin slices. Arrange on a plate with tomato slices. Sprinkle with remaining ingredients. Serves 4.

South Pacific Mixed Salad

3 cups shredded lettuce	1 cup thinly sliced raw carrots
1 medium-sized cucumber, peeled and sliced	1 avocado, peeled and cubed
	¼ cup drained capers
3 medium-sized tomatoes, peeled and quartered	1 large green pepper, cleaned and cut into strips
2 cups cooked green peas	About ¾ cup French dressing

Combine ingredients, except dressing, and toss in a salad bowl. Just before serving, moisten with French or any other favorite dressing. Serves 4 to 6.

Venezuelan Spinach-Avocado Salad

½ pound fresh spinach	1 garlic clove, crushed
1 large onion, peeled and thinly sliced	Salt and pepper to taste
	Juice of ½ lemon
2 hard-cooked eggs, peeled and chopped	1 tablespoon vinegar
	¼ cup olive or vegetable oil
1 large ripe avocado, peeled and cubed	

Wash spinach and cut off stems. Dry. Tear into bite-size pieces. Put in a salad bowl. Separate onion into rings; add with remaining ingredients to spinach. Toss lightly. Serves 6.

Caribbean Vegetable-Fruit Salad

A good buffet salad.

- 3 cups shredded green cabbage
- 2 medium-sized red apples, unpeeled, cored and chopped
- 1 cup diced raw or canned pineapple
- 1 large tomato, peeled and chopped
- 1 green pepper, cleaned and chopped
- 1 medium-sized ripe avocado, peeled and cubed
- About $2/3$ cup mayonnaise
- $1/4$ cup pineapple juice
- 1 tablespoon fresh lime or lemon juice

Combine vegetables and fruit in a large bowl. Mix together the mayonnaise, pineapple and lemon juices, and stir into vegetable-fruit mixture. Serves 8 to 10.

Iowa Hot Potato Salad

A good salad for an informal outdoor meal.

- 6 medium-sized potatoes
- Boiling salted water
- 4 slices thin bacon
- 1 medium-sized onion, peeled and chopped
- $1/4$ cup vegetable oil
- 2 tablespoons vinegar
- 1 teaspoon powdered dry mustard
- 2 teaspoons sugar
- Salt and pepper to taste
- 3 tablespoons chopped fresh parsley
- 2 hard-cooked eggs, peeled and sliced

Scrub potatoes and cook with skins on in boiling salted water until tender. Drain, peel, and cut into thin slices while still hot. Meanwhile, fry bacon in a skillet until crisp. Drain on absorbent paper and chop. Reserve. Using 2 tablespoons of heated bacon fat, sauté onions until tender. Add to potatoes. Combine oil, vinegar, mustard, sugar, salt and pepper, and heat. Pour over potatoes. Stir in parsley and mix well. Serve garnished with eggs. Serves 6.

Swedish Vegetable-Macaroni Salad

1 cup macaroni elbows	1½ cups light cream or milk
5 tablespoons milk	¼ cup prepared horseradish
1 cup cooked green peas	2 tablespoons fresh lemon juice
1 cup sliced cooked carrots	1 teaspoon sugar
½ cup cooked cut-up green beans	Salt and pepper to taste
1 tablespoon butter or margarine	2 tablespoons chopped fresh dill or parsley
1 tablespoon all-purpose flour	

Cook macaroni according to package directions until just tender; drain. Turn into a bowl. Add milk and vegetables. Mix well and chill 30 minutes. Melt butter in a saucepan. Stir in flour. Cook 1 minute. Gradually add cream and cook over low heat, stirring, until smooth and thick. Add horseradish, lemon juice, sugar, salt and pepper. Remove from heat. Cool. Mix with macaroni and vegetables. Serve garnished with dill or parsley. Serves 4 to 6.

Burmese Green Bean Salad

The Burmese prepare and serve bean sprouts, cabbage, and cauliflower in the same way as these green beans.

¾ pound fresh green beans or 1 package (9 ounces) frozen cut-up green beans	4 tablespoons sesame or vegetable oil
Boiling salted water	2 teaspoons turmeric powder
2 large onions, peeled	1 teaspoon sugar
2 garlic cloves, minced	Salt and pepper to taste
	3 tablespoons cider vinegar

Cook fresh or frozen green beans in a little boiling salted water until just tender. Drain. Cut onions into fine shreds and sauté with garlic in 3 tablespoons oil until soft. Add turmeric powder, sugar, and pepper and cook 1 minute. Heat vinegar in a small saucepan. Add 1 tablespoon oil. Pour over cooked beans. Top with sautéed onion mixture. Serves 4.

Chinese Broccoli Salad

2 pounds fresh broccoli or 2 packages (10 ounces each) frozen broccoli spears	1 garlic clove, crushed
	2 tablespoons soy sauce
	About 2 tablespoons vinegar
Boiling salted water	About ¼ cup peanut or vegetable oil
2 cups canned bean sprouts, drained	
	½ teaspoon sugar
1 cup sliced radishes	Salt and pepper to taste
½ cup sliced bamboo shoots	
½ cup chopped scallions, with tops	

Cut off and discard any wilted leaves and some of stalk of fresh broccoli. Steam, covered, in a little boiling salted water until just tender, or cook frozen broccoli according to package directions. Drain and cut into bite-size pieces. Combine with remaining ingredients, using enough vinegar and oil to moisten vegetables. Serves 6 to 8.

Indonesian Vegetable Salad with Peanut-Butter Sauce

This traditional salad, called *gado gado*, is served with a spicy peanut sauce over a combination of raw and cooked vegetables. It is an attractive and different dish for a party.

1 cup cooked cut-up green beans	1 medium-sized cucumber, peeled and sliced
1 cup sliced cooked carrots	⅓ cup peanut butter
1 cup sliced boiled potatoes	1 garlic clove, crushed
1 cup shredded green cabbage	½ to 1 teaspoon cayenne
1 large tomato, peeled and sliced	½ teaspoon sugar
	1 cup milk

Arrange vegetables attractively on a platter. Refrigerate. Heat peanut butter in a small saucepan until runny. Add remaining ingredients and cook over low heat, stirring, until smooth. Cool. Serve with vegetables. Serves 6.

Curious Vegetables

There is such a diverse selection of vegetables grown throughout the world that it would be impossible for most persons to know them all. In fact, the majority of Americans are familiar with very few vegetables. Among the most interesting of the lesser-known varieties are those that could well be called curious, odd or unusual.

There is no particular definition of a curious vegetable. In such a large country as the United States a vegetable that is commonplace to a person living in the Southwest would be unusual in New England, and vice versa. Perhaps a vegetable could be considered curious when its appearance is strange, even unattractive, or its name is odd-sounding. Some of the vegetables in this section were once popular on our forefathers' tables but have been ignored—even forgotten—in modern times.

One of the great delights in cooking is to "discover" a new variety of food and add it to the family repertoire. Fortunately, a wide selection of vegetables is now available in our markets and stores. It is no longer necessary to search out those once sold only in Oriental or South American specialty food shops, for many of them are now displayed alongside completely familiar ones. We have only to purchase them and learn how to prepare them.

As Americans travel more frequently, they also increase

their knowledge of unusual vegetables and will begin to search for them in their own home markets. It may occasion some surprise to discover that what has been considered a "new" vegetable is really a very ancient one, which has been neglected and deserves a renaissance.

This section of recipes is for cooks who wish to explore and enjoy the curious vegetables. Unfortunately, it has not been possible to include all these wonderful strange foods, but here, at least, is a selection for a number of these choice but all too often overlooked vegetables.

Chinese Snow Peas and Mushrooms

The snow pea, also called sugar pea and Chinese pea pod, has a flat pod that is so tender that the entire vegetable—both pea and pod—is eaten. The pea has a bright green color and a firm, crisp pod, and is sold both fresh and frozen. The ends and the strings, if any, must be removed, and the peas should be used as promptly as possible so they retain their freshness. Snow peas are widely used in Chinese cookery—added to soups, meat, and poultry dishes, and cooked with other vegetables.

- 1 pound fresh snow peas
- 1 pound fresh mushrooms
- 3 tablespoons peanut or vegetable oil
- 6 scallions, cleaned and cut into 1-inch pieces
- 1 small piece ginger root, minced
- 1 to 2 tablespoons soy sauce

Cut ends and any strings from snow peas, and rinse well in cold water. Wipe dry and refrigerate until ready to cook. Wash mushrooms quickly or wipe with wet paper toweling to remove any dirt. Cut off any tough stem ends. Cut lengthwise from caps through stems. Heat oil in a wok or skillet. Add scallions and stir-fry 1 minute. Add mushrooms and stir-fry 2 minutes over fairly high heat. Then add peas and ginger root, and stir-fry a minute or so, until peas are tender but still crisp. Add soy sauce and serve at once. Serves 4.

Note: Frozen snow peas may be used as a substitute, but they must first be thoroughly defrosted and wiped dry.

Oriental Snow Peas with Celery

1 pound fresh snow peas	2 teaspoons cornstarch
2 tablespoons peanut or vegetable oil	2 tablespoons cold water
	2 tablespoons soy sauce
¼ cup sliced scallions, with tops	1 teaspoon sugar
4 large stalks celery, cleaned and cut into 2-inch slices	Pepper to taste

Cut ends and strings, if any, off snow peas. Rinse and wipe dry. Heat oil in a wok or skillet. Add scallions, celery, and snow peas and stir-fry over fairly high heat for 2 minutes. Combine remaining ingredients and add at once. Cook, stirring, 1 or more minutes until peas are tender but a little crisp. Serve at once. Serves 4.

Cardoons in Tomato Sauce Italiano

The cardoon, a thistlelike plant related to the globe artichoke but similar in appearance to a bunch of celery, has a tough outer skin and stalks that must be removed before cooking as only the inner stalks are edible. Cardoons are generally blanched before being eaten. Once cut, they must be rubbed with lemon juice or put in acidulated water to prevent discoloring. Highly prized by the French and the Italians, cardoons can be blanched and then served with simply a vinaigrette dressing, or dipped into batter and deep-fried, or added to a salad.

2 bunches cardoons	1 teaspoon dried oregano
Boiling salted water	½ teaspoon dried basil
5 tablespoons butter or margarine	Salt and pepper to taste
	1 cup fine dry bread crumbs
½ cup chopped mushrooms	½ cup grated Parmesan cheese
3 cups tomato sauce	

Remove tough outer and wilted stalks of cardoons. Take off all leaves and strings from inner stalks and cut into 3-inch pieces. Trim heart and cut into pieces. Drop at once into cold water with a little vinegar or lemon juice added to it. Cook in boiling salted water, covered, until just tender. Drain well. Melt 2 tablespoons butter in a saucepan. Add mushrooms and sauté 3 minutes. Add tomato sauce, oregano, basil, salt and pepper, and cook slowly, uncovered, 5 minutes. Pour some of sauce into a shallow baking dish. Arrange cardoon pieces in the dish and cover with remaining sauce. Sprinkle top with bread crumbs and cheese. Melt remaining 3 tablespoons butter and sprinkle over top. Bake in preheated 375° F. oven until hot and bubbly, about 20 minutes. Serves 6.

Cardoons Polonaise

A topping that is commonly served over such cooked vegetables as asparagus, Brussels sprouts, cauliflower, and cardoons is simply a combination of butter and bread crumbs that may or may not include other foods. It is called *polonaise*.

2 bunches cardoons, washed, cleaned, and cooked as in previous recipe	3 hard-cooked eggs, shelled and chopped
½ cup butter	3 tablespoons chopped fresh parsley
½ cup fine dry bread crumbs	

Place hot cooked cardoons on a platter. Melt butter in a skillet over medium heat and brown lightly. Add bread crumbs and brown them also lightly but do not burn. Pour over cardoons at once and sprinkle with egg and parsley. Serves 4 to 6.

Dasheens

The dasheen, a plant of the taro family grown for its large tubers, is a staple food of the Southeast Pacific and Africa. In the Orient it is known as the Chinese potato and is cooked and served the way the white potato is in many parts of the world. Dasheens were introduced into Florida from the West Indies. The flavor of these vegetables is nutlike, something like a combination of potato and chestnut. Because dasheens have less moisture than the potatoes, they are mealy when cooked. They have a dark-brown skin that should be well scrubbed but need not be peeled.

Dasheens are best boiled or baked in their skins. Boiled ones can be peeled, cut in slices, and sautéed in butter, or mashed and seasoned with butter, salt, pepper, and parsley.

In Florida there is a dasheen salad, similar to potato salad, made with chopped celery, onion, well-seasoned mayonnaise, and chopped cooked cold dasheens.

Hawaiian Bok Choy Medley

Bok choy, one of the best and most widely used Chinese vegetables, is commonly called Chinese cabbage or chard. The name, however, means "white cabbage." The vegetable has long thick white stalks and large dark green leaves with white veins. Most recipes call for the use of only the stalks, but the leaves can be used in soup. Bok choy, which has a very pleasant flavor and crispness, should be used promptly and never overcooked. It is sold fresh by the bunch the year round in Oriental or specialty food stores and in some supermarkets.

1 medium-sized head fresh bok choy	1 piece ginger root, minced
3 tablespoons peanut or vegetable oil	2 cups pineapple chunks
	1 large green pepper, cleaned and chopped
1 bunch scallions, cleaned and cut into 1-inch pieces, with tops	¼ cup pineapple juice
	½ cup sliced water chestnuts
	Salt and pepper to taste

Rinse bok choy and cut off stem ends. Cut out any wilted parts and remove green leaves from stems. Wipe dry and cut into bite-size pieces. Heat oil in a large skillet. Add scallions and sauté 1 minute. Add bok choy and sauté, turning often, about 3 minutes. Add remaining ingredients and cook, stirring often, about 7 minutes, until bok choy is just tender but crisp. Serves 6.

Chinese Stir-Fry Bok Choy and Vegetables

A good buffet or party dish.

1 tablespoon cornstarch	3 tablespoons peanut or vegetable oil
3 tablespoons cold water	
1 tablespoon vinegar	2 medium-sized onions, peeled and sliced thickly
2 tablespoons soy sauce	
1 tablespoon sugar	1 cup thinly sliced scraped carrots
1 medium-sized head fresh bok choy	
	1 cup sliced bamboo shoots
1 medium-sized head romaine	Pepper to taste

Combine cornstarch, water, vinegar, soy sauce, and sugar in a small pitcher and set aside. Rinse bok choy and cut off stem end. Cut any wilted parts and remove green leaves from stems. Wipe dry and cut into bite-size pieces. Wash and clean romaine and tear into small pieces; wipe dry. Heat oil in an extra-large skillet and add onions and carrots. Stir-fry about 3 minutes. Add bok choy and stir-fry about 4 minutes. Add romaine and bamboo shoots and stir-fry 1 or 2 minutes. Pour in soy-sauce mixture and season with pepper. Stir-fry, mixing often, 4 or 5 minutes. Serves about 10.

Creamed Oyster Plant

Oyster plant, also called salsify, is an herb cultivated for its dark tapering root, which is prepared and served as a vegetable. The name derives from the flavor of its white flesh, which has a resemblance to that of an oyster, but it is also thought to be like that of a globe artichoke. Oyster plant must be scraped or peeled and kept in acidulated water until cooked. Primarily a winter vegetable, it was a great favorite of our forefathers, who generally boiled the flesh and served it simply with butter or a sauce, or made it into fritters.

2 bunches oyster plant
2 tablespoons vinegar or lemon juice
2 tablespoons butter or margarine
2 tablespoons all-purpose flour
1 cup light cream or milk
Salt and pepper to taste
2 tablespoons chopped fresh parsley

Wash and scrape oyster plant. Cut into small pieces and drop into cold water with vinegar or lemon juice until ready to cook. Boil in small amount of unsalted water until tender, about 30 minutes. Drain. Melt butter in a saucepan. Stir in flour and cook 1 minute. Gradually add cream, or milk, and cook slowly, covered, until smooth and thick. Season with salt and pepper. Pour over warm cooked oyster plant and sprinkle with parsley. Serves 4.

French Sautéed Oyster Plant

- 2 bunches oyster plant
- 2 tablespoons vinegar or lemon juice
- Boiling water
- 3 tablespoons olive or vegetable oil
- 2 garlic cloves, crushed
- 1 medium-sized onion, peeled and chopped
- 2 large ripe tomatoes, peeled and chopped
- Salt and pepper to taste
- ¼ cup chopped fresh parsley

Wash and scrape oyster plant. Cut into small pieces and drop in cold water with vinegar or lemon juice while preparing. Boil in small amount of unsalted water until tender, about 30 minutes. Drain. Heat oil in a skillet. Add garlic and onion, and sauté until tender. Add tomatoes and cook 1 minute. Mix in oyster plant, salt and pepper, and sauté 2 or 3 minutes. Serve sprinkled with parsley. Serves 4.

German Celeriac In Cheese Sauce *

Celeriac, also known as celery root and celery knob, is a member of the celery family and is widely used in Europe cooked as a hot vegetable or raw as a salad. It is a good, inexpensive vegetable with a dark, gnarled whiskery root and leafy tops. Only the root is eaten and small ones are best. The vegetable has a celery-like flavor and can be cooked in interesting ways.

* Another recipe for celeriac is on page 146 (in Salads).

- 1½ pounds celeriac
- Boiling salted water
- 5 tablespoons butter or margarine
- 3 tablespoons all-purpose flour
- 1½ cups light cream or milk
- ½ cup grated Parmesan cheese
- Salt and pepper to taste
- About ¼ cup fine dry bread crumbs

Wash celeriac well and cut off tops. Peel roots and cube. Cook in a little boiling salted water until tender, about 20 minutes. Drain, mash, and place in a greased baking dish. Meanwhile, melt 3 tablespoons butter in a saucepan. Mix in flour and cook 1 minute. Gradually add cream or milk and cook slowly, stirring, until thickened. Add cheese and leave over low heat until melted. Season with salt and pepper. Pour over mashed celeriac and sprinkle with bread crumbs. Bake in preheated 375° F. oven until golden and browned on top, about 20 minutes. Serves 4.

French Sautéed Celeriac

- 1 large onion, peeled and chopped
- 1 garlic clove, crushed
- About 3 tablespoons butter or margarine
- 2 large ripe tomatoes, peeled and sliced
- 3 cups shredded cleaned celeriac
- Salt and pepper to taste
- 2 tablespoons chopped fresh basil or parsley

Sauté onion and garlic in heated butter in a skillet until tender. Add tomatoes and sauté 1 minute. Add celeriac and 2 tablespoons water. Cook slowly, covered, until tender, about 10 minutes. Remove cover. Season with salt and pepper, and cook until most of liquid has evaporated. Sprinkle with basil or parsley. Serves 4.

Caribbean Calabaza

Sometimes called a West Indian pumpkin, the calabaza is a squash of varying size, shape, and skin coloring, but usually large with a deep yellow or orange flesh and a delicate flavor. Calabaza resembles the Hubbard or butternut squash which can be used as a substitute for it, if desired. The vegetable is generally boiled and served as an accompaniment to meats or poultry, added to soups and stews, and used in making breads.

- 1½ pounds calabaza
- Boiling salted water
- 2 tablespoons vegetable oil
- 2 garlic cloves, crushed
- 1 large onion, peeled and minced
- 1 green pepper, cleaned and minced
- 1 teaspoon ground hot red peppers
- 1 cup tomato juice
- ½ teaspoon dried basil or oregano

Peel calabaza and remove any seeds and fibers. Cut into cubes and cook in a little boiling salted water until tender. Drain and set aside and keep warm. Heat oil in a saucepan. Add garlic and onion and sauté until tender. Add remaining ingredients and cook, uncovered, 10 minutes. Pour over warm cooked calabaza. Serves 4 to 6.

Flemish Endive Salad

An imported vegetable that is considered a luxury food is the small snow-white compact endive that is also called Belgian or French endive, *witloof* or *chicorée*. Actually it is a type of chicory that was discovered by accident in Brussels, Belgium, in 1845 when a gardener, experimenting to improve chicory roots (sometimes used as a substitute for coffee), found that a group of them, left in a cellar for several days, began sprouting firm leaves shading from white to yellow. Further testing and planting led to the development of the heads, grown underground in covered trenches, and over the years this pampered vegetable has developed into an important Belgian industry. For some strange reason endive thrives primarily in that country, but it is also cultivated by some hobbyists in America. Endive is delicate and has a slightly bitter taste. Although primarily used uncooked in salads, alone or with other foods, endive is delicious when cooked, either braised or baked; it can also be stuffed or made into soup.

2 heads endive	¼ cup vegetable oil
1 cup cooked green beans, cut French style	2 tablespoons wine vinegar
	2 teaspoons sugar
1 cup shoe-string cooked or canned beets	¼ teaspoon powdered mustard
	Salt and pepper to taste
1 medium-sized onion, peeled and chopped	¼ cup chopped fresh parsley

Trim the base of each endive. Discard any wilted outer leaves. Wash well under running water to remove all dirt. Wipe dry and cut in slices. Combine with remaining ingredients and chill before serving. Serves 4 to 6.

Belgian Baked Endive Mornay

12 heads endive	
Boiling water	3 tablespoons all-purpose flour
2 tablespoons fresh lemon juice	2 cups light cream or milk
3 tablespoons butter or margarine	1 cup shredded Gruyère or Swiss cheese
1 small onion, peeled and minced	Salt, freshly ground pepper, grated nutmeg, to taste

Trim the base of each endive. Discard any outer wilted leaves. Spread apart leaves and hold under running water to remove all dirt. Invert to drain. Put in a large saucepan and cover with boiling water. Add lemon juice and cook over low heat for 10 minutes. Take out and drain. Arrange in a shallow buttered baking dish. Melt butter in a medium-sized saucepan and sauté onion until tender. Mix in flour and cook 1 minute. Slowly add cream, or milk, and cook over low heat, stirring, until thick and smooth. Add cheese, salt, pepper, and nutmeg. Continue to cook over low heat, stirring occasionally, until cheese has melted. Pour over endive. Bake in preheated 400 F. oven until tender, about 25 minutes. Serves 6.

Jicama Salad

Jicama, pronounced *hee-kah-mah,* is a brown root vegetable that looks like a turnip but tastes like a fresh water chestnut. It has crisp white meat something like the potato—in fact, it is used in Mexico very much the way the potato is in the United States. Generally quite large, the vegetable is sold whole and cut into halves or quarters. It is also eaten uncooked in Mexico: served with chili powder and pieces of lemon as an appetizer and accompaniment for tequila drinks, and sold by street vendors as a snack. Uncooked jicama is also used in salad. To cook, wash well and boil, steam, or fry. Now sold in the United States in Latin American or Oriental markets, the vegetable is also called yam bean. Jicama is often used as a substitute for the water chestnut and the bamboo shoot as it is more economical.

- 2 cups peeled and chopped raw jicama
- 1 medium-sized onion, peeled and chopped
- 1 green pepper, peeled and chopped
- 1 cup diced cucumber
- 1/4 cup olive or vegetable oil
- 2 tablespoons fresh lime or lemon juice
- Salt and pepper to taste
- 2 tablespoons chopped fresh parsley

Combine ingredients in a bowl. Serves 4 to 6.

Japanese White Radish with Sesame

A long tapered white radish, called by the Japanese a *daikon,* is one of the most commonly used vegetables in the Orient. It resembles a Western radish or white turnip in flavor but is less sharp. Usually about 6 inches long but often of giant size, quite lengthy and heavy, the radish can be eaten raw, pickled, or cooked in a variety of ways. The Japanese treasure the radish not only as a vegetable but also when grated, shredded, or cut into fanciful shapes as an attractive garnish for soup and other dishes.

1 pound white radish (daikon), washed and sliced	1 tablespoon vinegar
	1 teaspoon sugar
1 cup thinly sliced raw carrots	2 tablespoons soy sauce
½ cup thinly sliced mushrooms	¼ cup toasted sesame seeds *

Combine radishes, carrots and mushrooms in a bowl and mix together the remaining ingredients. Add to vegetables. Serves 4.

* To toast sesame seeds, spread in a pan and toast in preheated moderate oven for 10 to 15 minutes, until golden brown.

Oriental White Radish Medley

2 cups sliced cleaned white radishes	1 teaspoon cornstarch
	2 tablespoons cold water
1 cup sliced scraped carrots	2 tablespoons soy sauce
1 cup green peas	Pepper to taste
1 cup chicken broth	

Combine vegetables and chicken broth in a saucepan and cook until tender, about 8 minutes. Combine remaining ingredients and add. Cook slowly, stirring, until a little thickened. Serves 4.

Puerto Rican Creamed Chayote

A member of the squash family that is highly esteemed in Caribbean and South American cuisines, the green or white pear-shaped chayote is actually a fruit but is prepared and served as a vegetable. Its flesh is firm and rather crisp. If the skin is not tough, it does not have to be peeled. Because the chayote has a bland flavor, it is generally seasoned with herbs or cheese. Chayote can be simply steamed and served like squash, but it is also good when cooked and marinated in a dressing, with or without other fruits and vegetables, to be served as an appetizer or salad. Also known as vegetable pear and *chrisotophine*, the chayote is available in many areas from November to April.

2 medium chayotes, washed	1 can (6 ounces) tomato paste
Boiling salted water	¼ teaspoon dried oregano
3 tablespoons butter or margarine	Salt and pepper to taste
3 tablespoons all-purpose flour	3 tablespoons chopped fresh parsley
1½ cups milk	

Cut chayotes crosswise through seeds into ½-inch slices or cubes. Put in 1 inch boiling salted water in a saucepan and cook slowly, covered, until tender about 12 minutes. Remove from heat; drain and keep warm. Melt butter in a saucepan. Add flour and cook 1 minute. Gradually add milk and cook slowly, covered, until thick and smooth. Stir in tomato paste, oregano, salt and pepper. Cook 5 minutes. Add chayote and parsley and leave on stove long enough to heat through. Serves 4.

Virgin Island Stuffed Chayotes

4 medium-sized chayotes	Salt and pepper to taste
Boiling salted water	About ⅓ cup fine dry bread crumbs
1 medium-sized onion, peeled and minced	3 tablespoons butter or margarine
1 cup grated Edam, Gouda, or other yellow cheese	

Scrub chayotes well and cut in halves lengthwise. Steam in a small amount of salted water until just tender. Drain and with a spoon carefully take out pulp and seeds, leaving a firm shell about ¼ inch thick. Mash pulp and seeds, and drain off any liquid. Add onion and cheese, and season with salt and pepper. Spoon filling into shells. Sprinkle with bread crumbs and dot with butter. Arrange in a shallow baking dish. Pour in about ½ inch water. Bake, uncovered, in a preheated 350° F. oven about 20 minutes, until top is browned and mixture is heated through. Serves 4.

Stir-Fried Celery Cabbage

Celery cabbage—also called nappa, long cabbage, and Chinese cabbage—grows like celery and is similar in appearance. The bunch has long and large tightly packed greenish-white crisp leaves that are strong veined and crinkled. Their flavor is appealing and refreshing. Originally from China, this cabbage is now widely available in America and may be prepared and cooked in many of the same ways as green cabbage. Raw, it is an excellent addition to salads, and it may also be added to soup and to meat and vegetable dishes. Another name for it is *pe-tsai*.

  About 1/3 cup peanut or vegetable oil
4 cups cut-up cleaned celery cabbage
1 cup thinly sliced scraped carrots
1 medium-sized green pepper, cleaned and sliced
1/4 cup water chestnut slices
1 garlic clove, crushed
3 tablespoons vinegar
2 tablespoons soy sauce
1 teaspoon sugar

Heat the oil in a large skillet and add the vegetables, one third of them at a time. After each addition, stir-fry 1 minute, then continue to stir-fry a few more minutes until tender but a little crisp. Add remaining ingredients, mix well, and leave on stove long enough to heat through. Serves 4.

Celery Cabbage with Sour Cream Dressing

2 tablespoons butter or margarine
1 medium-sized head celery cabbage, cleaned and cut into bite-size pieces
1 cup sour cream at room temperature
1 garlic clove, crushed
1/4 cup minced green onions, with tops
2 tablespoons chopped fresh parsley
Salt and pepper to taste

Melt butter in a large skillet. Add cabbage and sauté over fairly high heat until it begins to wilt. Meanwhile, combine remaining ingredients and spoon over cabbage. Leave over low heat, stirring occasionally, until heated through. Serves 4.

Hearts of Palm and Avocado Salad

Heart of palm, the white core of the young palm tree, is called cabbage palm or swamp cabbage in Florida where it is available fresh. Elsewhere it is sold in cans in specialty food stores and some supermarkets. The hearts look like cabbage and are eaten as a vegetable. They have a bland and delicate flavor and taste something like chestnuts. The hearts are commonly used in salads; they can also be cooked and made into such a dish as the recipe that follows.

- 1 can (14 ounces) hearts of palm, drained and cut into ¼ inch slices
- 1 medium-sized avocado, peeled and diced
- 1 small head lettuce, washed, drained, and broken
- 1 large green pepper, cleaned and sliced thinly
- About ½ cup Italian or French dressing
- Salt, pepper to taste

Combine all ingredients and toss lightly. Chill. Serves 4 to 6.

Hearts of Palm Gratinée

- 1 can (14 ounces) hearts of palm, drained and sliced lengthwise
- 2 tablespoons butter or margarine
- 2 tablespoons all-purpose flour
- 1 cup milk
- ⅔ cup diced sharp Cheddar cheese
- ¼ teaspoon powdered mustard
- Dash cayenne
- Salt and pepper to taste

Arrange hearts of palm in a shallow baking dish. Melt butter in a saucepan. Stir in flour and cook 1 minute. Gradually add milk, stirring as adding, and cook until thickened and smooth. Add cheese, mustard, cayenne, salt and pepper, and leave on stove until cheese melts. Spoon over hearts and put under the broiler for a minute or two, until hot and bubbly. Serves 4.

Breads & Desserts

Vegetables are not commonly associated with the ingredients used in breads and desserts, yet they can contribute greatly to their flavor and attractiveness. Furthermore, some vegetables considerably enhance the texture of baked goods.

The culinary roster of breads and desserts that can be made with vegetables is an extensive one. As a mainstay of the daily diet, yeast and quick breads, rolls and muffins, should be as appealing and good as possible. Certainly vegetables contribute to these factors although such foods are rarely served on our tables, and the same is true of cakes and cookies made with vegetables.

Better known are pies made with vegetables, particularly those using pumpkin, sweet potato, and squash, but tomatoes and zucchini, among others, can be equally appealing in pie. Vegetable puddings are age-old favorites, and innovative modern variations of them are a pleasure to prepare.

The following recipes represent some cosmopolitan favorites.

Colonial Squash Gems

Early American cooks made small muffins, called gems, that were baked in heavy cast-iron gem pans with shallow oblong divisions and rounded bottoms. Muffin pans can be used as a substitute. Cooked squash, or pumpkin, was often used in baked breads as it added flavor and enriched the texture.

1½ cups all-purpose flour	½ cup milk
¼ cup sugar	½ cup mashed cooked or canned squash
¾ teaspoon salt	
2 teaspoons baking powder	2 tablespoons melted butter or margarine
1 egg, beaten	

Sift together flour, sugar, salt, and baking powder into a large bowl. Add egg and milk, and mix well. Stir in squash. Spoon into greased muffin pans, filling each about two thirds full. Bake in preheated 350° F. oven about 25 minutes, or until tester inserted into them comes out clean. Makes about 10.

Tomato-Herb Whole Wheat Bread

This is a nutritious dark bread that has a delicious herb flavor.

1 package active dry yeast or 1 cake compressed yeast	(basil, rosemary, thyme, sage, savory, marjoram, and/or parsley)
¼ cup lukewarm water	
2 cups lukewarm tomato juice	About 5 tablespoons melted shortening
2 tablespoons sugar	
2 teaspoons salt	About 5½ cups sifted whole wheat flour
1 tablespoon mixed dried herbs	

Sprinkle or crumble yeast into lukewarm water in a large bowl. Leave a minute or two and stir to dissolve. Add tomato juice, sugar, salt, herbs, and 2 tablespoons shortening and mix well. Add 2 cups flour; beat well. Mix in remaining flour, enough to make a stiff dough; stir thoroughly. Turn out on a floured board and knead about 10 minutes, until smooth and elastic. Form into a large ball and put in a warm greased bowl. Brush with melted shortening. Cover with a clean light cloth and let rise in a warm place until doubled in bulk. Punch down and turn out on a floured board. Shape into two loaves and put in two greased 9- by 5- by 3-inch loaf pans. Brush tops with melted shortening. Cover and let rise until doubled in bulk. Bake in a preheated 375° F. oven about 40 minutes, until golden on top and done. Turn out on a rack and cool. Makes 2 loaves.

German Parsley-Potato Dumplings

2 cups boiled potatoes, cooled and riced
2 eggs
1/2 teaspoon salt
About 1 cup all-purpose flour
Dash pepper
2 tablespoons finely chopped fresh parsley
About 2 tablespoons melted butter
About 1/4 cup fine dry bread crumbs

Combine potatoes, eggs, and salt in a bowl. Beat well. Add enough flour to make a stiff dough, pepper, and parsley, and mix well. Shape into balls about 1½ inches in diameter. Drop into rapidly boiling salted water and cook until dumplings rise to top, about 12 minutes. Test by tearing one dumpling apart with two forks. Remove with a slotted spoon and drain. Serve at once. Garnish with melted butter and fine dry bread crumbs, mixed together, if desired. Serves 4 to 6.

Iowa Corn Fritters

These fritters may be served as a substitute for bread or, covered with maple syrup, as a supper dish.

1¾ cups all-purpose flour	½ cup milk
2 teaspoons baking powder	1 tablespoon melted shortening
1 teaspoon salt	2 cups cut fresh or canned whole-kernel corn
Dash pepper	
1 egg, slightly beaten	Fat for deep frying

Sift dry ingredients into a bowl. Combine egg, milk, and shortening, and pour into flour mixture. Stir until just smooth. Add corn and stir well. Drop by tablespoons into hot deep fat (375° F. on frying thermometer) and fry until golden on all sides, 3 to 5 minutes. Drain on absorbent paper. Serve at once. Serves 4.

Pumpkin-Walnut Bread

2 cups sifted all-purpose flour	½ cup milk
2 teaspoons baking powder	1 cup sugar
½ teaspoon baking soda	1 cup mashed cooked or canned pumpkin
1 teaspoon salt	
½ teaspoon ground nutmeg	¼ cup melted butter or margarine
1 teaspoon ground cinnamon	
2 eggs, well beaten	1 cup chopped walnuts

Sift flour, baking powder and soda, salt, and spices into a bowl. Combine eggs, milk, sugar, pumpkin, and butter in another bowl; mix well. Add sifted dry ingredients and mix until well blended. Fold in nuts. Turn into a greased 9- by 5- by 3-inch loaf pan. Bake in preheated 350° F. oven about 35 minutes, or until tester inserted into center comes out clean. Remove from stove and let rest a few minutes. Turn on a rack and cool. Serves about 10.

Onion-Cheese Bread

Serve warm with butter or use for making sandwiches.

1 package active dry yeast or 1 cake compressed yeast	1½ teaspoons salt
¼ cup lukewarm water	2 cups shredded Cheddar cheese
1½ cups skim milk	¾ cup chopped onion
2 tablespoons shortening	About 5 cups sifted all-purpose flour
2 tablespoons sugar	

Sprinkle dry yeast or crumble cake yeast into water in a large bowl. Let stand a few minutes. Stir to dissolve. Combine milk, shortening, sugar, salt, and cheese in a small saucepan and heat to lukewarm. Pour over yeast and mix well. Add onion and 2 cups flour; mix well. Add enough flour to make a stiff dough, one that will not stick to sides of bowl. Turn over. Cover and let rise in a warm place until doubled in bulk. Punch down and shape into 2 loaves. Place in 2 greased 9- by 5- by 3-inch loaf pans. Let rise, covered, until doubled in bulk. Bake in preheated 350° F. oven about 45 minutes, until golden on top and cooked through. Makes 2 loaves.

Parsley Biscuits

An easy-to-prepare and colorful quick bread.

2 cups sifted all-purpose flour	1 cup chopped fresh parsley, washed and dried
3 teaspoons baking powder	About 1 tablespoon melted butter or margarine
1 teaspoon salt	
¼ cup shortening	
⅔ cup milk	

Sift flour, baking powder, and salt into a bowl and cut in shortening with two knives or a pastry blender. Add milk and mix to form a soft dough. Turn out on a floured board and knead lightly. Roll out to a ½-inch thickness and cut with floured round cookie cutter into 2-inch biscuits. Brush tops with melted butter.

Arrange on cookie sheet and bake in preheated 450° F. oven 12 to 15 minutes, until tester inserted into center comes out clean and biscuits are golden. Makes about 14.

Soy and Whole Wheat Muffins

This is a nutritious bread made with two of our finest flours—soy, milled from soybeans, and whole wheat, milled from cleaned whole wheat grain.

- 1 cup soy flour
- 1 cup whole wheat flour
- 1 teaspoon salt
- 3 teaspoons baking powder
- 2 tablespoons sugar
- 3/4 cup milk
- 1 egg, beaten
- 2 tablespoons melted butter or margarine

Sift first five ingredients into a large bowl. Combine milk, egg, and butter, and add to dry ingredients. Mix just enough to combine ingredients. Spoon into greased muffin pans, filling two thirds full, and bake in a preheated 425° F. oven for 12 to 15 minutes. Makes 12 muffins.

Georgia Sweet-Potato Rolls

Sweet potatoes add a pleasing and distinctive flavor to yeast breads such as these rolls.

- 3/4 cup scalded milk
- 1/4 cup butter or margarine
- 1/4 cup light brown sugar
- 1/2 teaspoon salt
- 3/4 cup mashed cooked sweet potatoes
- 1 package active dry yeast or 1 cake compressed yeast
- 1/4 cup lukewarm water
- About 3 1/3 cups sifted all-purpose flour

Combine first five ingredients in a large bowl. Let stand until lukewarm. Sprinkle or crumble yeast into lukewarm water. Stir to dissolve. Add to sweet-potato mixture. Mix in enough flour to make a soft dough. Stir to thoroughly combine. Form into a ball and put in a greased warm bowl; turn over. Cover with a light cloth and let rise in a warm place until doubled in bulk. Punch down and let rise again until doubled in bulk. Form dough into small balls, and place 3 balls in each section of a well-greased muffin pan. Let rise, covered, until doubled in bulk. Bake in preheated 425° F. oven about 12 minutes, or until cooked. Makes about 3 dozen.

Carrot Halva from India

Halva, made with milk, flavorings, and a vegetable such as carrots or pumpkin, is a traditional dessert in India.

- 1 pound, about 8 medium-sized carrots
- 4 cups milk
- 2 tablespoons honey
- $1/2$ cup sugar
- 4 whole cardamoms, removed from pods and crushed
- $1/2$ teaspoon ground cinnamon
- 2 tablespoons butter or margarine
- $1/4$ cup seedless raisins (optional)
- Garnishes: About $1/3$ cup chopped pistachios and $1/3$ cup silver sprinkles

Scrape and coarsely grate carrots. Combine with milk in a large saucepan. Bring to a boil. Reduce heat and cook over moderate heat 1 hour, or until liquid is reduced and mixture is thick. Add honey, sugar, cardamoms, and cinnamon, and continue cooking about 15 minutes longer, until further thickened. Stir in butter and raisins and mix well. Turn out onto a plate and shape into a mound. Sprinkle with garnishes. Serve either warm or cold. Serves 6 to 8.

Chocolate Sauerkraut Cake

Although the principal ingredients seem a strange combination, the cake is a pleasant surprise and is certain to be a conversation piece.

½ cup butter or margarine	1 teaspoon baking soda
1 cup sugar	½ teaspoon salt
2 eggs	½ cup cocoa
1 teaspoon vanilla	1 cup water
2 cups sifted all-purpose flour	⅔ cup rinsed, drained, chopped sauerkraut
3 teaspoons baking powder	

Cream butter or margarine in a large bowl. Add sugar and eggs, and beat until light and fluffy. Mix in vanilla. Sift in flour, baking powder and soda, salt, and cocoa, adding alternately with water. Beat well and stir in sauerkraut. Mix again. Turn into a greased rectangular 13- by 9- by 2-inch baking dish and bake in a preheated 350° F. oven for about 40 minutes, until a tester inserted into the cake comes out clean. Remove from stove and let stand a few minutes. Invert onto a cake rack and cool. Frost with white icing, if desired. Serves about 12.

Maryland Sweet Potato Custard

An interesting variation of the traditional custard.

¼ cup white or brown sugar	1¾ cups milk
½ teaspoon salt	2 eggs, beaten
¼ teaspoon grated nutmeg	2½ cups grated raw sweet potato
½ teaspoon ground cinnamon	
1 teaspoon grated orange or lemon rind	1 tablespoon melted butter or margarine

Combine sugar, salt, nutmeg, cinnamon, orange or lemon rind, milk, and eggs in a bowl. Add sweet potatoes and butter, and spoon into six greased custard cups. Set in a pan filled with about 1 inch of hot water. Bake in a preheated 350° F. oven about 35 minutes, or until custard is set and a tester inserted into it comes out clean. Serve warm or cold. Serves 6.

Turkish Pumpkin-Nut Pudding

4 pounds pumpkin	1/2 cup ground walnuts
1 1/2 to 2 cups sugar	1 cup heavy cream, whipped
About 1/2 cup water	

Pare pumpkin and cut into 1-inch slices. Put in a saucepan, sprinkling sugar between layers. Add water. Cook, covered, stirring occasionally, over medium heat until pumpkin is tender and water is absorbed, about 1 hour. Add a little more water during cooking, if desired. Remove from heat. Mash and cool. Serve garnished with nuts and cream. Serves 4 to 6.

Carrot-Nut Cake

A good dessert for a buffet or holiday dinner.

1 1/2 cups peanut or vegetable oil	2 teaspoons baking soda
2 cups sugar	2 teaspoons ground cinnamon
4 eggs, beaten	1/2 teaspoon salt
3 cups all-purpose flour	2 cups finely grated raw carrots
3 teaspoons baking powder	1 cup chopped walnuts

Combine first three ingredients in a bowl. Sift in flour, baking powder and soda, cinnamon, and salt. Mix well. Add carrots and nuts, and combine thoroughly. Turn into a greased 10- by 4-inch tube pan and cook in a preheated 350° F. oven about 50 minutes, or until tester inserted into center comes out clean. Remove from stove and let stand a few minutes. Turn out on a rack and cool. Serves about 12.

Hoosier Green Tomato Pie

Indiana is proud of its superb tomatoes, and Hoosier cooks have created many interesting ways of using them. This is one of their good tomato desserts.

6 medium-sized green tomatoes
2 tablespoons fresh lemon juice
1 teaspoon grated lemon rind
½ teaspoon salt
¼ teaspoon ground cinnamon
¾ cup sugar
2 tablespoons cornstarch
2 tablespoons butter or margarine
Standard pastry for 2-crust 9-inch pie, prepared

Wash tomatoes and remove stem ends. Cut into thin slices. Combine sliced tomatoes, lemon juice and rind, salt, and cinnamon in a saucepan. Cook 10 minutes, stirring frequently. Mix sugar and cornstarch; add to tomato mixture. Cook until thickened and clear, stirring constantly. Add butter. Cool 1 minute. Spoon into 9-inch pie plate lined with pastry. Cover with top layer of pastry. Seal and flute edges. Cut a few slits in the top. Bake in preheated 375° F. oven about 40 minutes, until pie filling is cooked. Serves 6.

Old-Fashioned Squash Pie

Pumpkin may be used as a substitute for the squash, if desired

½ cup white sugar
½ cup brown sugar
½ teaspoon ground cinnamon
¼ teaspoon ground nutmeg
1½ cups mashed cooked or canned squash
½ teaspoon salt
2 eggs, beaten
1½ cups light cream or milk
1 9-inch unbaked pie shell

Combine first six ingredients in a bowl. Mix in eggs and then cream or milk. Turn into pie shell. Bake in preheated 450° F. oven for 15 minutes. Lower heat to 350° F. and bake another 35 minutes, or until tip of knife inserted into filling comes out clean. Serves 6.

Southern Sweet Potato and Nut Cake

This traditional Southern cake is an excellent dessert for summer meals, particularly those served outdoors.

½ cup butter or margarine, softened	¼ teaspoon baking soda
1 cup sugar	½ teaspoon salt
2 eggs	½ teaspoon ground cinnamon
1 cup mashed cooked sweet potatoes	½ teaspoon ground nutmeg
2 cups all-purpose flour	¼ teaspoon ground cloves
2 teaspoons baking powder	½ cup milk
	¾ cup chopped nuts

Cream butter, or margarine, in a large bowl. Add sugar and mix well. Add eggs, one at a time, and beat until light and fluffy. Stir in sweet potatoes and mix well. Sift together flour, baking powder and soda, salt, and spices, and add alternately with milk to sweet potato mixture. Stir in nuts and mix well. Turn into a greased loaf pan, 9- by 5- by 3-inches, and bake in a preheated 350° F. oven about 55 minutes, or until tester inserted into cake comes out clean. Remove from stove and let rest a few minutes. Turn out on a rack and cool. Serves about 10.

Puerto Rican Sweet Potato Pudding

3 cups mashed, cooked, or canned sweet potatoes	2 eggs, separated
½ cup brown sugar	1 teaspoon ground cinnamon
2 tablespoons all-purpose flour	½ teaspoon ground allspice
½ cup orange juice	¼ teaspoon salt
2 tablespoons melted butter or margarine	

Combine sweet potatoes, sugar, flour, orange juice, butter, and egg yolks in a large bowl. Mix in cinnamon, allspice, and salt. Beat egg whites until stiff and fold into sweet potato mixture. Spoon into a greased baking dish and cook in preheated 375° F. oven until puffed and browned, about 30 minutes. Serves 4.

Tomato-Soup Spice Cake

This old-time favorite made with condensed tomato soup may be served with or without white or lemon frosting.

½ cup butter or margarine	1 teaspoon ground cinnamon
1⅓ cups sugar	1½ teaspoons ground allspice
2 eggs	½ teaspoon ground cloves
2 cups sifted all-purpose flour	1 can (10¾ ounces) condensed tomato soup
4 teaspoons baking powder	
1 teaspoon baking soda	

Cream butter in a large bowl. Add sugar and beat until light and fluffy. Add eggs, one at a time beating after each addition. Sift in dry ingredients, adding alternately with soup. Combine well and turn into a greased rectangular 13- by 9- by 2-inch baking dish. Bake in preheated 350° F. oven about 40 minutes, or until a tester inserted into cake comes out clean. Remove from stove and let rest a few minutes. Turn out on a rack and cool. Serves about 12.

Berkshire County Potato-Custard Pie

An unusual version of custard pie.

1 medium-sized potato, peeled, boiled, and mashed	2 eggs, separated
2 tablespoons butter or margarine	½ cup milk
	Juice and grated rind of ½ lemon
¾ cup sugar	1 unbaked 9-inch pastry shell
⅛ teaspoon salt	

Combine potato, while still warm, with butter, sugar, and salt in a bowl. Beat until creamy and cool. Add egg yolks, milk, lemon juice, and rind; blend well. Beat egg whites until stiff and fold into potato mixture. Turn into pastry shell and bake in preheated 400° F. oven about 25 minutes, until filling is set and tester inserted into it comes out clean. Serves 6.

Costa-Rican Beet Pudding

An innovative colorful dessert that is simple to prepare.

1 cup beet juice	4 tablespoons cornstarch
1 cup water	1 tablespoon fresh lemon juice
1/2 cup sugar	About 2 tablespoons grated coconut
6 whole cloves	
1/4 teaspoon each of ground allspice and cinnamon	

Combine beet juice, 3/4 cup water, sugar, cloves, allspice, and cinnamon in a medium-sized saucepan. Cook slowly, uncovered, for 10 minutes. Mix cornstarch with remaining 1/4 cup water and stir into beet mixture. Cook slowly, stirring, until thickened, about 1 minute. Remove from stove. Take out and discard cloves. Add lemon juice; mix well. Spoon into dessert dishes or a bowl. Garnish with coconut. Serves 4.

New Hampshire Pumpkin Cake

A good holiday dessert.

1/3 cup butter or margarine	1 1/4 cups whole wheat flour
1 cup brown sugar	3 teaspoons baking powder
1/2 cup buttermilk	1/2 teaspoon baking soda
2 tablespoons dark molasses	1/2 teaspoon salt
2 eggs, well beaten	1/2 teaspoon ground nutmeg
2/3 cup mashed, cooked or canned pumpkin	1/2 teaspoon ground cinnamon
	1/2 teaspoon ground allspice

Cream butter or margarine in a large bowl. Mix in sugar and beat until light and fluffy. Combine buttermilk and molasses; add to creamed mixture and stir well. Add eggs and pumpkin. Stir in remaining ingredients and stir well. Turn into greased shallow 8-inch square baking dish and cook in preheated 375° F. oven about 25 minutes, until a tester inserted into center comes out clean. Serves 6 to 8.

Louisiana Yam-Coconut Pie

A variety of sweet potato developed in Louisiana has been misnamed the yam. The true yam belongs to a different plant family, is large, and has a whitish- to reddish-colored skin. The flesh can be white, yellow, or sometimes red but is generally deep orange with a high sugar content. In cookery yams and sweet potatoes can be used interchangeably.

1/2 cup brown sugar	2 eggs, slightly beaten
1 teaspoon ground cinnamon	1 cup milk
1/2 teaspoon ground allspice	2 tablespoons melted butter or margarine
1/4 teaspoon ground nutmeg	
1/2 teaspoon salt	1/3 cup shredded coconut
1 1/2 cups mashed cooked or canned yams	9-inch unbaked pastry shell

Combine sugar, spices, and salt in a bowl. Stir in yams. Combine eggs, milk and butter and add to yam mixture. Spoon into pastry shell. Bake in preheated 400° F. oven about 40 minutes, or until filling is set and a knife inserted into it comes out clean. Sprinkle with coconut and bake 5 minutes longer. Serves 6.

Pickles, Relishes, & Sauces

In the realm of vegetable cookery the possibilities for making pickles, relishes, and sauces are endless. Many are used to accent the flavors of other foods, but in some countries they also supply basic nutrients.

Although many pickles, relishes, and condiments, such as ketchup, are prepared in quantity and canned for later use, some of these selections can be made and eaten in a very short time. Some are cooked; others are not.

The blending of vegetables with spices, sugar, and vinegar or other ingredients gives interesting texture and flavor to vegetables that have long been treasured on the world's tables. There are any number of old-time favorites.

Many homemakers like to prepare their pickles and relishes when garden vegetables are at their prime and in abundant supply. Select vegetables that are fresh, firm, unbruised, and not too ripe. Wash thoroughly and use as soon as possible after gathering from the garden or purchasing from a market.

The following recipes reflect the worldwide diversity of these dishes; they will add zest and variety to whatever foods they complement.

Persian Pickled Eggplant

Serve with shish kebab, hamburgers, or lamb.

1 medium-sized eggplant, about 1¼ pounds	8 cardamom seeds, removed from pods and crushed
1 tablespoon powdered mustard	1 tablespoon turmeric powder
2 teaspoons celery seed	Salt and pepper to taste
½ teaspoon crushed red pepper	About ⅓ cup vinegar
2 garlic cloves, crushed	

Wash eggplant; remove stem ends; make several small slits in skin. Bake in preheated 400° F. oven until tender, about 45 minutes. Remove from stove and cool slightly. Turn into a bowl and chop. Drain off any liquid. Add remaining ingredients, except vinegar, and mix well. Add enough vinegar to moisten ingredients. Leave at room temperature for several hours, to blend flavors, or for 2 days in refrigerator. Serves about 6.

Sauerkraut Relish

A good relish to serve with pork or game.

1 pound sauerkraut, washed, drained, and chopped fine	1 medium-sized onion, peeled and minced
1 medium-sized green pepper, cleaned and chopped fine	½ cup chopped scraped carrots
1 medium-sized sweet red pepper, cleaned and chopped fine	1 teaspoon mustard seed
	Salt and pepper to taste
	About ⅓ cup sugar

Combine ingredients and mix well. Leave at room temperature for 24 hours before using, or refrigerate for 2 or 3 days. Makes 4 cups.

Indonesian Sweet-Sour Mixed-Vegetable Relish

A good relish to serve with lamb or pork.

- 1 cup julienne-cut scraped carrots
- 1 cup shredded green cabbage
- 1 cup drained bean sprouts
- Salt
- 1 large onion, peeled and cut into shreds
- 2 to 3 garlic cloves
- About $1/3$ cup vinegar
- 1 tablespoon sugar
- 1 to 2 teaspoons crushed hot red peppers

Put carrots, cabbage, and bean sprouts in a bowl and sprinkle with salt. Let stand 30 minutes and drain. Combine onion, garlic, vinegar, and sugar in a saucepan. Bring to a boil. Reduce heat and simmer 5 minutes. Remove from heat and add drained vegetables and peppers. Mix well and cool. Serves 6 to 8.

Green Tomato Relish

A very popular American condiment that is sometimes called piccalilli.

- 1 quart chopped, cored green tomatoes
- 1 quart chopped green cabbage
- 1 cup chopped sweet red peppers
- 1 cup chopped green peppers
- $1^{1}/_{2}$ cups chopped onions
- $1/3$ cup salt
- 2 cups firmly packed brown sugar
- 3 cups cider vinegar
- 2 tablespoons whole pickling spice

Put vegetables and salt in a large bowl and let stand overnight. Drain well. Press to remove all liquid. Combine sugar, vinegar, and spices, previously put in a small piece of clean cheesecloth and tied, in a saucepan and simmer 15 minutes. Add vegetables and bring to a boil. Remove and discard spice bag. Spoon into hot sterilized pint jars, leaving $1/2$ inch space at top. Adjust caps. Process in hot boiling water 5 minutes. Cool. Makes 4 pints.

Ceylonese Cauliflower Sambal

The Southeast Asian condiment called sambal, which is eaten with rice and curry, is generally made of a combination of vegetables and fruits mixed with seasonings. It can also be served with meats and poultry.

3 cups cut-up raw cauliflower	2 tablespoons turmeric powder
1 medium-sized carrot, scraped and sliced	1 tablespoon minced ginger root
1 medium-sized green pepper, cleaned and chopped	1 tablespoon sugar
½ cup chopped green onions	Salt and pepper to taste
1 medium tart apple, peeled, cored, and chopped	About ¼ cup vinegar

Combine all ingredients in a bowl, adding enough vinegar to moisten them, and leave to marinate, turning occasionally, in the refrigerator for 2 days. Serves about 6.

Dilled Green Beans

Dill, an aromatic herb, has feathery leaves, called dillweed, that are used fresh or dried as a popular seasoning, and seeds that are frequently used for pickling. Serve these beans as an appetizer or pickle.

2 pounds, about 2 quarts, green beans	½ teaspoon dill seed per pint jar
¼ teaspoon crushed red pepper per pint jar	1 clove garlic per pint jar
½ teaspoon whole mustard seed per pint jar	2½ cups cider vinegar
	2½ cups water
	¼ cup salt

Wash beans; drain. Cut off ends and cut into 1½-inch lengths. Pack into clean, hot sterilized jars. Add pepper, mustard and dill seeds, and garlic. Combine vinegar, water, and salt and heat to boiling. Pour over beans, filling to ½ inch of top of jar. Adjust lids. Process in boiling water about 5 minutes. Makes about 3½ pints.

California Zucchini Relish

Californians are devoted to zucchini and prepare it in every possible way. The best zucchini for this relish are small ones that are very tender.

> 8 cups finely chopped small tender zucchini
> 2 cups finely chopped onion
> ¼ cup salt
> About 1½ cups vinegar
> 1 cup sugar
> 2 teaspoons mustard seed
> 2 teaspoons celery seed
> 1 or 2 garlic cloves, crushed

Combine zucchini, onion, and salt in a bowl. Cover with water. Let stand overnight and drain well. Combine remaining ingredients and bring to a boil. Add drained zucchini and onion and bring to a boil again. Spoon into hot sterilized jars and seal. Makes about 4 pints.

Indian Rayta

This refreshing combination of yogurt and vegetables is generally served as a cooling contrast to hot curries. It is also refreshing served as an appetizer or salad.

> 2 cups plain yogurt
> ½ cup minced onions
> 1 cup minced raw carrots or other vegetables
> 1 to 2 teaspoons minced green or red chili peppers
> 1 teaspoon ground cumin powder
> Salt and pepper to taste
> ¼ cup finely chopped fresh coriander or parsley

Combine ingredients and chill. Serves 4 to 6.

Pennsylvania Dutch Corn Relish

This is one of the sweets and sours, homemade relishes and condiments, that are traditional staples of Pennsylvania Dutch cooking. It is a very good accompaniment to pork or other meats.

- 4 cups cut fresh corn, from about 9 ears
- 1 cup diced sweet red peppers
- 1 cup diced sweet green peppers
- 1 cup chopped onion
- 2 cups chopped green cabbage
- 1 tablespoon powdered dry mustard
- 2 teaspoons ground turmeric powder
- 2 teaspoons celery seed
- 1 teaspoon mustard seed
- 3/4 cup sugar
- 2 cups cider vinegar
- 2 teaspoons salt

To prepare corn, remove husks and silk from ears. Cook in salted boiling water 5 minutes. Cut corn from cobs with a sharp knife. Combine with remaining ingredients in a large saucepan or kettle and cook slowly, covered, 25 minutes. Pack in hot sterilized jars leaving 1/2 inch of top space. Adjust caps. Process in boiling water for 15 minutes. Makes about 3 1/2 pints.

Note: Frozen corn can be used as a substitute for fresh corn. Defrost thoroughly overnight in refrigerator or 2 to 3 hours at room temperature

Pakistan Tomato Chutney

Chutney, a well-seasoned Eastern relish, is generally made of fruits and spices and served with curry. This sweet relish can also be served with meats, fish, or vegetables.

- 2 pounds (about 8 medium-sized) ripe tomatoes, peeled and chopped
- 1 cup sugar
- 2 garlic cloves, crushed
- 2 to 3 teaspoons chili powder or crushed red pepper
- 2 teaspoons chopped ginger root
- Salt to taste
- 1/2 cup seedless raisins
- 1 cup vinegar

Combine all ingredients in a saucepan. Bring to a boil. Reduce heat and simmer, uncovered, stirring often, until thickened, about 45 minutes. Cool and spoon into jars. Makes about 1 pint.

Korean Cabbage Kimchi

Keem-chee as it is pronounced, a staple nutritious pickle valued for its nutrients, is made in Korea in considerable quantity each autumn and stored in large earthenware crocks that are buried in the ground. The pickle is fiery hot with red chili peppers and garlic. This is a milder version.

- 2 pounds green or celery cabbage
- 2 tablespoons salt
- 1/2 cup sliced scallions
- 1 or 2 garlic cloves
- 1 teaspoon sliced ginger root
- 1 to 3 teaspoons crushed or ground red peppers

Wash cabbage and cut into strips, discarding any core. Put in a large bowl and sprinkle with salt. Leave about 2 hours and drain, pressing to release all liquid. Combine with remaining ingredients and leave to marinate in a large bowl or glass jars for 2 to 3 days. Makes about 2 pints.

New Jersey Mixed-Pepper Relish

An easy-to-prepare relish that can be kept in the refrigerator several days and used as needed.

2 cups finely chopped green peppers	½ cup vinegar
2 cups finely chopped red peppers	½ cup sugar
1 medium-sized onion, peeled and minced	1 teaspoon salt

Combine peppers and onion in a bowl. Bring vinegar to a boil in a saucepan. Add sugar and salt. Pour over vegetables. Let stand 2 to 3 hours before serving. Or spoon into jars and keep, tightly covered, in refrigerator for about 1 week, if desired. Makes about 4 cups.

Japanese Spinach with Sesame

This Japanese vegetable "pickle" can be made also with cabbage or cauliflower, if desired.

1 pound fresh spinach	¼ cup soy sauce
¼ cup toasted sesame seeds *	Pepper to taste

Clean spinach by cutting off stems and removing any wilted leaves. Wash and put in a saucepan with water clinging to leaves. Cook until tender, about 8 minutes. Drain to remove all liquid. Chop and mix with sesame seeds, soy sauce, and pepper. Cool and serve cold. Serves 4.

* To toast sesame seeds put in a skillet and heat until they become dark colored.

Pickled Vegetables

  4 cups vinegar
  1 cup sugar
  1 tablespoon salt
  ½ cup mixed pickling spices, tied in small pieces of cheesecloth
  7 cups shredded green cabbage
  2 cups chopped onions
  3 cups chopped green peppers
  3 cups shredded scraped carrots

Combine vinegar, sugar, salt, and spices in a saucepan and bring to a boil. Boil 10 minutes. Add vegetables and boil 5 minutes. Remove and discard spices. Spoon mixture into 4 hot sterilized jars leaving ½ inch of space at top. Adjust lids and process in boiling water for 30 minutes. Makes 4 pints.

New England Sweet-Sour Beets

New Englanders prepare beets with a sweet-sour sauce and serve the dish as a type of relish or as an accompaniment to meats and fish.

  1 tablespoon cornstarch
  2 teaspoons sugar
  ¾ cup beet juice
  ¼ cup cider vinegar
  Salt and pepper to taste
  1 tablespoon butter or margarine
  2 cups diced cooked beets

Combine cornstarch, sugar, beet juice, vinegar, salt and pepper, in a saucepan. Heat, stirring, until thick and smooth. Mix in butter and beets, and cook, stirring, until heated through. Serves 4.

Ripe Tomato Marmalade

Although marmalade, usually a preserve of fruit, is frequently used as a spread for bread, it can also be served with meats or vegetables and used in cooking, particularly for sauces and glazes.

12 cups or 3 quarts peeled, sliced tomatoes	2 lemons
6 cups sugar	2 cups water
2 teaspoons salt	4 sticks cinnamon
2 oranges	2 teaspoons whole cloves

Combine tomatoes, sugar, and salt in a large saucepan. Peel oranges and lemons. Slice peel very thin; boil in water 5 minutes and drain. Slice orange and lemon and remove seeds. Add with peel to tomato mixture. Put spices in a small piece of cheesecloth and tie. Add to tomato mixture. Bring slowly to a boil; then cook over brisk heat, stirring constantly, for about 50 minutes, until thickened. Remove and discard spices. Pour marmalade into hot sterilized jars and seal. Makes about 5 pints.

Japanese Pickled Cucumbers

Pickles are important to Japanese meals as their flavor goes well with rice, fish, and other traditional dishes. Celery cabbage can be prepared in the same way. Sometimes hot red peppers are added to the ingredients listed below.

2 medium-sized cucumbers, peeled	2 to 3 tablespoons soy sauce
Salt	3 tablespoons white vinegar
	1 teaspoon sugar

Cut cucumbers in halves lengthwise. Slice thinly. Sprinkle with salt and let stand 15 minutes. Press to release all liquid. Add remaining ingredients and mix well. Serves 4 to 6.

Basic Tomato Sauce

A good sauce to serve with cooked pasta, stuffed vegetables, meat loaf, meat patties, or cornmeal dishes.

- 2 large onions, peeled and chopped
- 4 garlic cloves, crushed
- 3 tablespoons olive or vegetable oil
- 2 cans (6 ounces each) tomato paste
- 2 cups water
- 1 teaspoon crumbled dried basil
- 1 teaspoon crumbled dried oregano
- 2 bay leaves
- Salt and pepper to taste

Sauté onions and garlic in heated oil until tender. Add tomato paste; mix well. Stir in remaining ingredients. Cook slowly, uncovered, 30 minutes. Thin with more water while cooking, if desired. Remove and discard bay leaves. Makes about 4 cups.

Minted Onion-Cucumber Relish

Serve with lamb or vegetables.

- 1/2 cup vinegar
- 1 tablespoon sugar
- 3/4 cup chopped fresh mint
- 1 cup thinly sliced small white onions
- 1 cup chopped peeled cucumbers
- Salt and pepper to taste

Heat vinegar and sugar in a small saucepan. Add mint leaves and heat over a low fire for 15 minutes. Remove from stove and add onions, cucumbers, salt and pepper. Chill. Makes about 2 cups.

Austrian Horseradish Cream Sauce

A good sauce for cold cooked fish such as salmon.

½ cup heavy cream	1 teaspoon sugar
2 tablespoons freshly grated or prepared horseradish, drained	½ teaspoon white vinegar Pinch of salt

Whip cream until thick. Stir in remaining ingredients and mix well. Chill. Serve cold. Makes about 1 cup.

Fresh Mushroom Sauce

Serve with cooked pasta, meats, or vegetables.

4 scallions, with tops, sliced	1 cup vegetable or beef bouillon
¼ cup butter or margarine	1 cup sour cream at room temperature
2 cups sliced fresh mushrooms	
1 tablespoon fresh lemon juice	
Salt and pepper to taste	2 tablespoons minced dill or parsley
2 tablespoons all-purpose flour	

Sauté scallions in heated butter in a medium-sized saucepan until tender. Add mushrooms and lemon juice and sauté 4 minutes. Season with salt and pepper. Stir in flour and mix well. Gradually add bouillon, and cook slowly, stirring often, until thickened, about 5 minutes. Mix in sour cream and dill; leave over low heat long enough to heat through. Makes about 2 cups

Italian Green Sauce

A good sauce to serve with cooked pasta.

¼ cup olive or vegetable oil	1 garlic clove, minced
5 tablespoons butter or margarine	3 cups chopped fresh spinach
½ cup minced scallions with tops	⅓ cup chopped fresh parsley
	Salt and pepper to taste

Heat oil and butter in a medium-sized saucepan. Add scallions and garlic and sauté until tender. Add spinach and parsley, and cook, stirring, 1 minute. Season with salt and pepper. Remove from heat. Makes about 1 cup.

English Cucumber Sauce

A good sauce to serve with fish.

½ cup peeled chopped cucumbers	White pepper to taste
Salt	1 tablespoon white vinegar
2 tablespoons chopped scallions, with tops	1 cup sour cream

Sprinkle cucumbers with salt and let stand 30 minutes. Drain. Combine with remaining ingredients and mix well. Makes about 1½ cups.

Provençal Garlic Sauce

A member of the lily family, along with the onion, leek, scallion, chive, and shallot, garlic is a controversial food, treasured by those who know and appreciate it, and abhorred by those who do not. A compound bulb made up of small, thin, paper-covered cloves, garlic is available dried and is valued for its potent flavor and odor. When cooking with garlic, break off a clove from the bulb and remove outer skin. Chop, mince, or crush in a garlic press. This sauce, actually a garlic mayonnaise, is used in Provence as a dip for raw and cooked vegetables. It is called *aïoli*.

6 cloves garlic, peeled	About 1 cup olive oil
2 egg yolks, lightly beaten	1 tablespoon fresh lemon juice
Pinch of salt	

Pound garlic in a mortar with a pestle. Add egg yolks and salt. Beat well with a wooden spoon. Begin adding oil very slowly, drop by drop, beating steadily. When fairly thick, add remaining oil in a steady stream, and then the lemon juice. Season with pepper. Leave at room temperature. The consistency should be that of a thick mayonnaise. Makes about 1½ cups.

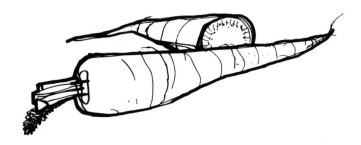

Old-Fashioned Tomato Ketchup

America's most popular condiment, ketchup or catsup, introduced by English colonists, is most often made with ripe red tomatoes. It can, however, also be prepared with mushrooms, walnuts, berries, and fruits as it was in the past.

7 pounds firm ripe red tomatoes	2 teaspoons celery seed
1/3 cup brown sugar	1/2 stick cinnamon
2 1/2 teaspoons salt	1 teaspoon whole black peppercorns
1 cup cider vinegar	
1 teaspoon dry mustard	

Wash, core, and cut tomatoes into quarters. Cook, uncovered, in a kettle for 30 minutes, stirring occasionally. Purée or put through a food mill. Add sugar, salt, vinegar, and mustard. Tie celery seed, cinnamon, and peppers in a piece of cheesecloth, and add to kettle. Cook about 1 hour longer, until thickened, stirring several times. Remove and discard spices. Pour into hot sterilized bottles and seal at once. Makes about 1 1/2 pints.

Index

Acorn Squash, Vermont, 134
aioli, 198
Alabama Turnip Greens and Pot Likker, 106
Albanian Minted Vegetable Casserole, 83
Algerian Baked Eggs with Vegetables, 80
Alsatian Sauerkraut, 134
American dishes: Alabama Turnip Greens and Pot Likker, 106; American Cassoulet, 124; Boston Baked Beans, 116; California Artichoke-Tomato Cocktail, 11; California Spinach-Zucchini Casserole, 88; California Zucchini Relish, 189; Collards, Southern, 105; Collards with Sour-Cream Dressing, 99; Colonial Squash Gems, 172; Connecticut Baked Jerusalem Artichokes, 136; Creole Tomato-Rice Gumbo, 35; Creole Wax Beans, 139; Down East Corn-Potato Chowder, 32; Down East Fiddleheads, 104; Early American Wilted Lettuce, 95; Florida Tuna-Stuffed Green Peppers, 39; Green Tomato Relish, 187; Hoosier Green Tomato Pie, 180; Idaho Stuffed Potatoes, 54; Iowa Corn Fritters, 174; Iowa Hot Potato Salad, 152;

American dishes (*cont.*)
Lima Beans, Texas Style, 131; Louisiana Red Beans, 114; Maryland Sweet Potato Custard, 178; Mormon Pioneer Greens, 96; Nebraska Stuffed Turnips, 45; New England Boiled Dinner, 65; New England Sweet-Sour Beets, 193; New Hampshire Pumpkin Cake, 183; New Hampshire Scalloped Potatoes and Parsnips, 139; New Jersey Mixed-Pepper Relish, 192; New Mexican Chiles Rellenos, 90; New Mexican Stuffed Peppers, 90; Oregon Salmon-Stuffed Tomatoes, 54; Ozark Sweet-Sour Cole Slaw, 144; Pennsylvania Dutch Corn Relish, 192; Pennsylvania Dutch Potato Pancakes, 89; Red Flannel Hash, 66; Refried Beans, 119; Shaker Greens with Eggs, 102; Southern Hoppin' John, 123; Southern Sweet Potato and Nut Cake, 181; Southwestern Frijoles, 118; Vermont Acorn Squash, 134; Vermont Mess o' Greens with Cornmeal Dumplings, 105; Virginia Cress Stew, 108
Anchovy-Stuffed Onions from Sicily, 46

Antipasto, Italian Vegetable, 15
appetizers, 1–18, 51, 113, 188, 189
Artichoke(s), 11; Italian Stuffed, 50; Moroccan, with Lemon-Olive Dressing, 133; -Tomato Cocktail, California, 11. *See also* Jerusalem Artichokes, Connecticut Baked.
Asparagus, 81; Canapés from Germany, 5; Creamed Fresh, Austrian Egg-Noodle Ring with, 71; Flemish, 81; German Suppe, 22; Poor Man's, 103
Australian Orange Carrots, 133
Austrian: Egg-Noodle Ring with Creamed Fresh Asparagus, 71; *Gurkensalat*, 146; Horseradish Cream Sauce, 196
Avocado: Salad, Hearts of Palm and, 170; Salad, Venezuelan Spinach-, 151

Bahmi Goreng, Indonesian, 73
Balkan Raw Vegetable-Yogurt Dip, 16
Bamboo Shoots, Stir-Fry Snow Peas with Mushrooms and, 127
Barley-Stuffed Cabbage Leaves, Polish, 47
Bean(s): Boston Baked, 116; Brazilian Black, 124; Chili, Dip from Mexico, 4; Creole Wax, 139; Dilled Green, 188; -Filled Enchiladas, Baked Corn-, 61; Hutspot, Dutch, 119; Indian Green, with Coconut, 129; Italian White, with Tunafish, 117; Louisiana Red, 114; Pasta Soup, Tuscan-, 32; Picnic Herbed White, 121; Quiche, Crookneck Squash-Green, 87; Refried, 119; Salad, Burmese Green, 153; Soup, Cuban Black, 112; Soup, Portuguese Kale-, 25; Turkish Cold Green, 132; Turkish Taverna, 116; Yugoslavian Cold White, 121. *See also* Egyptian Fool; Lima Beans.
Bean-Sprout: Appetizer, Oriental, 3; Salad, Oriental, 145
Beet(s), 33; Greens, Belgian Creamed, 99; New England Sweet-Sour, 193; Pudding, Costa-Rican, 183; Salad, Near Eastern, 146; Soup, Russian Cold, 33
Belgian: Baked Endive Mornay, 165; Creamed Beet Greens, 99; Fresh Mushroom-Herb Appetizer, 6
Berkshire County Potato-Custard Pie, 182
Biscuits, Parsley, 175
Black Beans(s): Brazilian, 124; Soup, Cuban, 112

Bohemian Creamed Kohlrabi, 136
Bok Choy, 159; Chinese Stir-Fry, and Vegetables, 160; Hawaiian, Medley, 159; Indonesian, 104
borscht, 33
Boston Baked Beans, 124
Brazilian Black Beans, 124
Bread(s), 171–176; Onion-Cheese, 175; Pumpkin-Walnut, 174; Tomato-Herb Whole Wheat, 172
British dishes, *see* English dishes; Scotch
Broccoli, 27; Indienne in Patty Shells, 67; Ring with Sour Cream Sauce, Summer, 89; Salad, Chinese, 154; Soup, Cold, 27
Brussels Sprouts, Flemish, with Cheese, 128
Bubble and Squeak from England, 138
Burmese Green Bean Salad, 153
Butternut Squash, Caribbean, 164
Butter-Steamed Lettuce with Cheese, 101

Cabbage, 41; Israeli Stuffed, 41; Kimchi, Korean, 191; Leaves, Polish Barley-Stuffed, 47. *See also* Celery Cabbage; Cole Slaw, Ozark Sweet-Sour; Sauerkraut.
Cake: Carrot-Nut, 179; Chocolate Sauerkraut, 178; New Hampshire Pumpkin, 183; Southern Sweet Potato and Nut, 181; Tomato-Soup Spice, 182
Calabaza, Caribbean, 164
California: Artichoke-Tomato Cocktail, 11; Spinach-Zucchini Casserole, 88; Zucchini Relish, 189
Canapés, Asparagus, from Germany, 5
Caponata, Sicilian, 17
Cardoons, 157; Polonaise, 158; in Tomato Sauce Italiano, 157
Caribbean dishes, 112; Caribbean Calabaza, 164; Caribbean Vegetable-Fruit Salad, 152; Cuban Black Bean Soup, 112; Puerto Rican Sweet Potato Pudding, 181; Puerto Rican Swiss Chard, 102; Virgin Islands Stuffed Chayotes, 168
Carrot(s), 4; Australian Orange, 133; Flemish, 130; soup, 29; Halva from India, 177; -Nut Cake, 179; Provençal Marinated, 4; -Sauerkraut Salad, 149; -Sweet Potato Tzimmes, Israeli, 140
Cassoulet, American, 124

Index

Cauliflower, 23; Garnished Whole, from Sicily, 91; -Potato Curry, Near Eastern, 81; Sambal, Ceylonese, 188; Soup, Cream of, Czech, 23; Salad, Turkish, 143
"Caviar," Russian Mushroom, 12
Celeriac, 162; French Sautéed, 163; German, in Cheese Sauce, 162; Rémoulade, French, 146
Celery: Oriental Snow Peas with, 157; Victor, 90
Ceylonese: Cauliflower Sambal, 188; Vegetable Curry, 69
Celery Cabbage: Indonesian, 104; Japanese Pickled, 194; with Sour Cream Dressing, 169; Stir-Fried, 169
chard, see Bok Choy; Swiss Chard
Chayote: Puerto Rican Creamed, 167; Virgin Island Stuffed, 168
Cheese: Butter-Steamed Lettuce with, 101; Flemish Brussels Sprouts with, 128; Onion-, Bread, 175; -Stuffed Cucumbers, 52. *See also* Cottage Cheese.
Chicken: -Olive-Filled Pimientos from Morocco, 44; -Vegetable Ragout, Parisian, 75
Chicory Gratinée, French, 98
Chick-Pea(s): Greek Baked, 120; Mexican, 122; Purée, Near Eastern, 113
Chiles Rellenos, New Mexican, 90
Chili Bean Dip from Mexico, 4
Chinese dishes: Chinese Broccoli Salad, 154; Chinese Egg Foo Yong, 78; Chinese Snow Peas and Mushrooms, 156; Chinese Stir-Fry Bok Choy and Vegetables, 160; Chinese Sub Gum Chow Mein, 71; Chinese Vegetable Fried Rice, 84; Hong Kong Spinach-Mushroom Soup, 27; Oriental Snow Peas with Celery, 157; Oriental Stir-Fry Greens, 100; Stir-Fried Celery Cabbage, 169; Stir-Fry Snow Peas with Mushrooms and Bamboo Shoots, 127
Chocolate Sauerkraut Cake, 178
Chop Chay, Korean, 58
Chow Mein, Chinese Sub Gum, 71
Chutney, Pakistan Tomato, 191
Circassian Vegetable-Yogurt Soup, 31
Coconut: Indian Green Beans with, 129; -Yam Pie, 184
Colache, Mexican, 79

Cole Slaw, Ozark Sweet-Sour, 144
Collards: with Sour Cream Dressing, 99; Southern, 105
Colonial Squash Gems, 172
Connecticut Baked Jerusalem Artichokes, 136
Corn, 79; -Bean-Filled Enchiladas, Baked, 61; -and-Chili Caldo, Mexican, 34; Iowa Fritters, 175; -Potato Chowder, Down East, 32; Pudding, South American, 138; Relish, Pennsylvania Dutch, 190; -Stuffed Pimientos Mexicana, 49; -Stuffed Zucchini, Mexican, 40
Corsican Stuffed Onions, 38
Costa-Rican Beet Pudding, 183
Cottage Cheese: Baked Mustard Greens and, 100; Peruvian Potato Salad with, 147; -Vegetable Dip, English, 13
Couscous, North African Vegetable, 57
Creole: Tomato-Rice Gumbo, 35; Wax Beans, 139
Cress Stew, Virginia, 108
Croquettes, French Green Pea, 6
Crookneck Squash, 43; -Green Bean Quiche, 87
crudités, 2
Cuban Black Bean Soup, 112
Cucumber(s), 3, 14; Boats from France, 14; Cheese-Stuffed, 52; Japanese Pickled, 194; Relish, Minted Onion-, 175; Sauce, English, 197; Tomato Salad Ring with, 150; Viennese Wilted, 146; -Yogurt Soup, Iranian Cold, 25
Curried Lentils from Pakistan, 123
Curry: Ceylonese Vegetable, 69; Green-Leaf, from India, 95; Near Eastern Potato-Cauliflower, 81
Custard: Maryland Sweet Potato, 178; Old-Fashioned Turnip, 130; Pie, Berkshire County, 81
Czech Cream of Cauliflower Soup, 23

Dal from India, 122
Dandelion Greens, Wilted, 97
Danish Vegetable Salad with Cheese, 142
Dasheens, 159
desserts, 171, 177–184
Dilled Green Beans, 188
dips, 4, 13, 16, 198
Down East: Fiddleheads, 104; Corn-**Potato Chowder**, 32

Dumplings: German Parsley-Potato, 173; Vermont Cornmeal, 106
Dutch Bean Hutsput, 119

Egg(s): Algerian Baked, with Vegetables, 80; Chinese, Foo Yong, 78; Creamed, with Spinach Ring and Walnuts, 83; -Noodle Ring with Creamed Fresh Asparagus, Austrian, 71; Omelette Basque, 77; Shaker, Greens with, 102
Eggplant, 3, 14; Appetizer, Riviera, 14; -Filled Green Peppers, South American, 51; French Mushroom-Stuffed, 42; Moussaka, 74; alla Parmigiana, 59; Persian Pickled, 186; Ratatouille of Provence, 67; Sicilian Caponata, 17; Turkish Lamb-Rice Stuffed, 44
Egypt, Fool from, 114
Enchiladas, Baked Corn-Bean-Filled, 61
Endive, 164; Belgian Baked, Mornay, 165; Salad, Flemish, 164
English dishes: Bubble and Squeak from England, 138; English Cottage Cheese-Vegetable Dip, 13; English Cucumber Sauce, 197
Esau's Lentil Pottage, 111

Fennel Salad, Italian, 150
Fiddleheads, Down East, 104
Finnish Summer Vegetable Soup, 34
Fish-Vegetable Plaki from Greece, 60
Flemish: Asparagus, 81; Brussels Sprouts with Cheese, 128; Carrots, 130; Endive Salad, 164
Florida Tuna-Stuffed Green Peppers, 39
Fool from Egypt, 114
French dishes, 2; Corsican Stuffed Onions, 38; *crudités*, 2; Cucumber **Boats from France**, 14; French Celeriac, Sautéed, 163; French Celeriac Rémoulade, 146; French Chicory Gratinée, 98; French Green Pea Croquettes, 6; French Mushroom-Stuffed Eggplant, 42; French Puréed Sorrel, 101; French Sautéed Oyster Plant, 162; Mushrooms Provençal, Stuffed, 52; Onion-Olive Pizza from Provence, 15; Parisian Watercress Soup, 107; Potage Crécy, 29; Potage Crème de Champignons, 24; Provençal Garlic Sauce, 198; Provençal Marinated Carrots, 4; Ratatouille of Provence, 67; Salad

French dishes (*cont.*)
Niçoise, 149; Soupe à l'Oignon Gratinée, 30; Soupe au Pistou, 28; Tomatoes Provençal, Baked, 135; Zucchini Quiche au Fromage, 7
French Canadian Split-Pea Soup, 112
Fried Rice, Chinese Vegetable, 84
Frijoles: *negros*, 124; *refritos*, 119; Southwestern, 118
Fritters, Iowa Corn, 174

Garlic Sauce, Provençal, 198
Gazpacho from Andalusia, 26
Georgia Sweet-Potato Rolls, 176
German dishes: Alsatian Sauerkraut, 134; Asparagus Canapés from Germany, 5; **German Asparagus Suppe, 22;** German Carrot-Sauerkraut Salad, 149; German Celeriac in Cheese Sauce, 162; German Hop Sprouts, 106; German Parsley-Potato Dumplings, 173; German Sauerkraut Balls, 9
Ghiveciu, Romanian, 62
Goulash, Hungarian Vegetable, 68
Grape Leaves: Greek Stuffed, with Lemon Sauce, 48; Persian Rice-Filled, 55
Greek: Baked Chick-Peas, 120; Eggplant Moussaka, 74; Fish-Vegetable Plaki from Greece, 60; Rice-Stuffed Tomatoes, 53; Salad, Classic, 145; Spinach Pie, 92; Stuffed Grape Leaves with Lemon Sauce, 48; Vegetable Medley, 135
Green: -Leaf Curry, Indian, 95; Sauce, Italian, 197; Tomatoes, Fried, with Milk Gravy, 128; Tomato Pie, Hoosier, 180; Tomato Relish, 187; -Vegetable Casserole, Persian, 77
Green Bean(s), 129; Dilled, 188; Indian, with Coconut, 129; Quiche, Crookneck Squash-, 87; Salad, Burmese, 153; Turkish Cold, 132
Greens, 93–108; Alabama Turnip, and Pot Likker, 106; Baked Mustard, and Cottage Cheese, 100; Belgian Creamed Beet, 99; Indonesian Mustard, 104; Mormon Pioneer, 96; Oriental Stir-Fry, 100; Shaker, with Eggs, 102; Vermont Mess o', 105; Wilted Dandelion, 97
Gumbo, Creole Tomato-Rice, 35

Halva from India, Carrot, 177

Hash, Red Flannel, 66
Hawaiian Bok Choy Medley, 159
Hearts of Palm: and Avocado Salad, 170; Gratinée, 170
Herb(s), 19–20; Appetizer, Belgian Fresh Mushroom, 6; -Tomato Whole Wheat Bread, 172
Hong Kong Spinach-Mushroom Soup, 27
Hoosier Green Tomato Pie, 180
Hoppin' John, Southern, 123
Hop Sprouts, German, 106
Horseradish Cream Sauce, Austrian, 196
Hungarian Vegetable Goulash, 68
Hutspot, Dutch Bean, 119

Idaho Potatoes, Stuffed, 54
Indian: Carrot Halva from India, 177; Dal from India, 122; Green Beans with Coconut, 129; Green-Leaf Curry, 95; Lentil-Rice Pot, 115; Rayta, 189; Vegetable Pachadi, 5; Vegetable Pilau, 82. *See also* Pakistan.
Indonesian: Bahmi Goreng, 73; Mustard Greens, 104; Sweet-Sour Mixed-Vegetable Relish, 187; Vegetable Salad with Peanut Butter Sauce, 154
Iowa: Corn Fritters, 174; Hot Potato Salad, 152
Iranian: Cold Cucumber-Yogurt Soup, 25; Vegetable Khoreshe, 58. *See also* Persian.
Iraqui Vegetable Pie, 85
Irish Leek and Potato Soup, 20
Israeli: Carrot-Sweet Potato Tzimmes, 140; Stuffed Cabbage, 41
Istanbul Vegetable Stew, 72
Italian dishes, 2; Anchovy-Stuffed Onions from Sicily, 46; Cardoons in Tomato Sauce Italiano, 157; Eggplant Parmigiano, 59; Garnished Whole Cauliflower from Sicily, 91; Italian Fennel Salad, 150; Italian Green Sauce, 197; Italian Stuffed Artichokes, 50; Italian Stuffed Mushrooms, 46; Italian Vegetable Antipasto, 15; Italian White Beans with Tuna Fish, 117; Macaroni-Zucchini Timballo, 86; Mediterranean Stuffed Zucchini, 48; Minestrone alla Romano, 36; Risi e Bisi from Venice, 131; Sicilian Caponata, 17; Tomato-Zucchini Lasagna, 63; Tuscan-Bean Pasta Soup, 32

Japanese: Pickled Cucumbers, 194; Spinach with Sesame, 192; Vegetable Sukiyaki, 62; Vegetable Tempura, 17; White Radishes with Sesame, 166
Jerusalem Artichokes, Connecticut Baked, 136
Jicama Salad, 166

Kale, 25; -Bean Soup, Portuguese, 25; and Oatmeal, Scotch, 94
Ketchup, Old-Fashioned Tomato, 199
Khoreshe from Iran, Vegetable, 58
Kidney Beans, Russian, 117
Kohlrabi, Bohemian Creamed, 136
Korean: Cabbage Kimchi, 191; Chop Choy, 58

Lamb-Rice Stuffed Eggplant, Turkish, 44
Lasagna, Tomato-Zucchini, 63
Latin American Vegetable Stew, 64
Lebanese Stuffed Squash with Yogurt Sauce, 43
Leek(s), 19; Marinated, Romanian Style, 10; and Potato Soup, Irish, 20; Scotch Braised, 132
Lemon Sauce, 49
Lentil(s): Curried, from Pakistan, 123; Dal from India, 122; Pottage, Esau's, 111; -Rice Pot, Indian, 115
Lettuce, 94, 141; Butter-Steamed, with Cheese, 101; Cream of, Soup, 24; Early American Wilted, 95
Lima Beans: Texas Style, 131; in Tomato Sauce, 120
Lithuanian Mixed-Vegetable Salad, 148
Louisiana: Red Beans, 114; Yam-Coconut Pie, 148

Macaroni: Swedish Vegetable-, Salad, 153; -Zucchini Timballo, Italian, 86
Marinated: Carrots Provençal, 4; Leeks, Romanian Style, 10
Maryland Sweet Potato Custard, 178
Mediterranean Stuffed Zucchini, 48
Mess o' Greens with Cornmeal Dumplings, Vermont, 105
Mexican dishes: Baked Corn-Bean-Filled Enchiladas, 61; Chili Bean Dip from Mexico, 4; Corn-Stuffed Pimientos Mexicana, 49; Jicama Salad, 166; Mexican Chick-Peas, 122; Mexican Colache, 79; Mexican Corn-and-Chili Caldo, 34; Mexican Corn-Stuffed Zucchini, 40

Middle Eastern dishes: Circassian Vegetable Yogurt Soup, 31; Egyptian Fool, 114; Esau's Lentil Pottage, 111; Iranian Cold Cucumber-Yogurt Soup, 25; Iranian Vegetable Khoreshe, 58; Iraqui Vegetable Pie, 85; Israeli Carrot-Sweet Potato Tzimmes, 140; Israeli Stuffed Cabbage, 41; Istanbul Vegetable Stew, 72; Lebanese Stuffed Squash with Yogurt Sauce, 43; Miniature Potato Balls from Turkey, 13; Persian Green-Leaf Vegetable Casserole, 77; Persian Pickled Eggplant, 186; Persian Rice-Filled Grape Leaves, 55; Spinach Borani, 3; Turkish Cauliflower Salad, 143; Turkish Cold Green Beans, 132; Turkish Lamb-Rice Stuffed Eggplant, 44; Turkish Pumpkin-Nut Pudding, 179; Turkish Taverna Beans, 116. *See also* Near Eastern.
Minestrone alla Romana, 36
Miniature Potato Balls from Turkey, 13
Minted Onion-Cucumber Relish, 195
Mormon Pioneer Greens, 96
Moroccan: Artichokes with Lemon-Olive Dressing, 133; Pimientos, Olive-Chicken-Filled, 44
Moussaka, Greek Eggplant, 74
Muffins: Colonial Squash Gems, 172; Soy and Whole Wheat, 176
Mushroom(s), 6; Chinese Snow Peas and, 156; -Filled Polish Pancakes, 70; -Herb Appetizer, Belgian Fresh, 6; Italian Stuffed, 46; Provençale Stuffed, 52; Sauce, Fresh, 196; Soufflé, 87; -Spinach Soup, Hong Kong, 27; Stir-Fry Snow Peas with, and Bamboo Shoots, 127; -Stuffed Eggplant, French 42
Mustard Greens: and Cottage Cheese, Baked, 100; Indonesian, 104

Near Eastern: Beet Salad, 146; Chick-Pea Purée, 113; Potato-Cauliflower Curry, 81. *See also* Greek; Middle Eastern dishes.
Nebraska Stuffed Turnips, 45
New England Boiled Dinner, 65. *For other New England recipes, see individual names under* American dishes.
New Hampshire: Pumpkin Cake, 183; Scalloped Potatoes and Parsnips, 139
New Jersy Mixed-Pepper Relish, 192

New Mexican Chiles Rellenos, 90
Noodle-Egg Ring with Creamed Fresh Asparagus, Austrian, 71
North African Vegetable Couscous, 57
Norwegian: Spiced Red Cabbage, 127; Summer Salad, 144
Nut: Cake, Carrot-, 179; Cake, Southern Sweet Potato and, 181

Omelette Basque, 77
Onion(s), 15, 19; Anchovy-Stuffed, from Sicily, 46; -Cheese Bread, 175; Corsican Stuffed, 38; -Cucumber Relish, Minted, 195; -Olive Pizza from Provence, 15; Soupe à l'Oignon Gratinée, 30; Tart, Swiss, 80
Oregon Salmon-Stuffed Tomatoes, 54
Oriental: Bean-Sprout Appetizer, 3; Bean-Sprout Salad, 145; Snow Peas with Celery, 157; Stir-Fry Greens, 100; White Radish Medley, 167. *See also* Chinese dishes; Indonesian; Korean; Japanese.
Oyster Plant, 161; Creamed, 161; French Sautéed, 162

Pachadi, Indian Vegetable, 5
Pakistan: Curried Lentils from, 123; Tomato Chutney, 191
Pancakes: Pennsylvania Dutch Potato, 89; Polish Mushroom-Filled, 70
Parisian Watercress Soup, 107
Parsley, 108; Biscuits, 108; Butter, 108; Fried, 108; -Potato Dumplings, German, 173
Parsnips, New Hampshire Scalloped Potatoes and, 139
Pea(s), 6; Croquettes, French Green, 6; Risi e Bisi from Venice, 131. *See also* Snow Peas.
Pennsylvania Dutch: Corn Relish, 190; Potato Pancakes, 89
Peppers, 39; Florida Tuna-Stuffed Green, 39; New Mexican Stuffed, 90; Relish, Mixed, 192; Soufflé, Red and Green, 85; South American Eggplant-Filled, 51
Persian: Green-Vegetable Casserole, 77; Pickled Eggplant, 186; Rice-Filled Grape Leaves, 55; Spinach Borani, 3
Peruvian: Mixed Vegetable Soup, 30; Potato Salad with Cottage Cheese, 147; Sweet-Potato Appetizers, 11

pickles, 185–186, 188, 191–194
Pie: Berkshire County Potato Custard, 182; Greek Spinach, 92; Hoosier Green Tomato, 180; Iraqui Vegetable, 85; Italian Macaroni-Zucchini, 86; Louisiana Yam-Coconut, 184; Old-Fashioned Squash, 180
Pilau, Indian Vegetable, 82
Pimientos, 44; Mexicana Corn-Stuffed, 40; Moroccan Olive-Chicken-Filled, 44
Pizza, Onion-Olive, from Provence, 15
Plake, Fish-Vegetable, from Greece, 60
Polish: Barley-Stuffed Cabbage Leaves, 47; Mushroom-Filled Pancakes, 70
Poor Man's Asparagus, 103
Portuguese Kale-Bean Soup, 25
Potage: Crécy, 29; Crème de Champignons, 24
Potato(es), 19; Balls, Miniature, from Turkey, 13; -Cauliflower Curry, Near Eastern, 81; Chowder, Down East Corn-, 32; -Custard Pie, Berkshire County, 182; Dumplings, German Parsley-, 173; Idaho, Stuffed, 54; and Leek Soup, Irish, 20; Pancakes, Pennsylvania Dutch, 89; and Parsnips, New Hampshire, 139; Roll, Russian, 78; Salad, Iowa Hot, 152; Salad, Peruvian, with Cottage Cheese, 147; Swedish Mashed Rutabaga and, 137. *See also* Sweet Potatoes.
Provençal(e): Baked Tomatoes, 135; Garlic Sauce, 198; Marinated Carrots, 4; Onion-Olive Pizza, 15; Ratatouille, 67; Salade Niçoise, 149; Soupe au Pistou, 28; Stuffed Mushrooms, 52
Pudding: Costa-Rican Beet, 183; Puerto Rican Sweet Potato, 181; South American Corn, 138; Turkish Pumpkin-Nut, 179
Puerto Rican: Sweet Potato Pudding, 181; Swiss Chard, 102
Pumpkin, 22; Cake, New Hampshire, 183; -Nut Pudding, Turkish, 179; Pie, Old-Fashioned, 180; Soup from Venezuela, 22; -Walnut Bread, 174

Quiche: Crookneck Squash-Green Bean, 87; Zucchini, au Fromage, 7

Radishes, 9–10; in Sour Cream, 9. *See also* White Radishes.

Ragout, Parisian Chicken-Vegetable, 75
Ratatouille of Provence, 67
Rayta, Indian, 189
Red Bean, Louisiana, 114
Red Flannel Hash, 66
Refried Beans, 119
relishes, 185–195
Rice: Black-Eyed Peas and, 123; Chinese Vegetable Fried, 84; -Filled Grape Leaves, Persian, 55; Pilau, Indian Vegetable, 82; Pot, Indian Lentil-, 115; Pot, Swedish Vegetable-, 65; Risi e Bisi from Venice, 131; -Stuffed Tomatoes, Greek, 53; -Tomato Gumbo, Creole, 35;
Risi e Bisi from Venice, 131
Riviera Eggplant Appetizer, 14
Rolls, Georgia Sweet Potato, 176
Romanian: Ghiveciu, 62; Marinated Leeks, 10
Russian: Cold Beet Soup with Sour Cream, 33; Kidney Beans, 117; Mushroom "Caviar," 12; Potato Roll, 78; Winter Vegetable Soup from Russia, 21; *zakuski*, 1
Rutabaga and Potatoes, Swedish Mashed, 137

Salad(s), 141–154: Burmese Green Bean, 153; Caribbean Vegetable-Fruit, 152; Celery Victor, 90; Chinese Broccoli, 154; Classic Greek, 145; Danish Vegetable, with Cheese, 142; dressing for, 142; Flemish Endive, 164; German Carrot- Sauerkraut, 149; Hearts of Palm and Avocado, 170; Indonesian Vegetable, with Peanut-Butter Sauce, 154; Iowa Hot Potato, 152; Italian Fennel, 150; Italian Tomato-Bread, 148; Lithuanian Mixed Vegetable, 148; Near Eastern Beet, 146; Norwegian Summer, 144; Oriental Bean-Sprout, 145; Ozark Sweet-Sour Cole Slaw, 144; Peruvian Potato, with Cottage Cheese, 147; Salade Niçoise, 149; South Pacific Mixed, 151; Swedish Vegetable-Macaroni, 153; Tomato, Ring with Cucumbers, 150; Tomato Flower, 143; Turkish Cauliflower, 143; Venezuelan Spinach-Avocado, 151; Viennese Wilted Cucumbers, 146
Salade Niçoise, 149

Salmon-Stuffed Tomatoes, Oregon, 54
Sambal, Ceylonese Cauliflower, 188
Sauce(s), 185; Austrian Horseradish, 196; Basic Tomato, 195; Cheese, 162; English Cucumber, 197; Fresh Mushroom, 196; Italian Green, 197; Lemon, 49; Peanut-Butter, 154; Provençal Garlic, 198; Sour Cream, 89; Yogurt, 43
Sauerkraut, Alsatian, 134; Balls, German, 9; Cake, Chocolate, 178; Relish, 186; Salad, German Carrot-, 149
Scallions, 103; Poor Man's Asparagus, 103
Scalloped Potatoes and Parsnips, New Hampshire, 139
Scandinavian dishes, 1; Danish Vegetable Salad with Cheese, 142; Norwegian Spiced Red Cabbage, 127; Summer Salad, 144; Swedish Mashed Rutabaga and Potatoes, 137; Swedish Vegetable-Macaroni Salad, 153; Swedish Vegetable-Rice Pot, 65
Scotch: Braised Leeks, 132; Kale and Oatmeal, 94
Shaker Greens with Eggs, 102
Sicilian Caponata, 17
Snow Peas, 156; Chinese, and Mushrooms, 156; Oriental, with Celery, 157; Stir-Fry, with Mushrooms and Bamboo Shoots, 127
Sorrel, French Puréed, 101
Soufflé: Fresh Mushroom, 87; Red and Green Pepper, 85
Soup(s), 19–36; Cold Broccoli, 27; Czech Cream of Cauliflower, 23; Gazpacho from Andalusia, 26; German Asparagus Suppe, 22; Hong Kong Spinach-Mushroom, 27; Iranian Cold Cucumber-Yogurt, 25; Irish Leek and Potato, 20; Minestrone alla Romana, 36; Parisian Watercress, 107; Portuguese Kale-Bean, 25; Potage Crécy, 29; Potage Crème de Champignons, 24; Pumpkin, from Venezuela, 22; Soupe à l'Oignon Gratinée, 30; Soupe au Pistou, 28; Spring, 29; Winter Vegetable, from Russia, 21
Soupe: à l'Oignon Gratinée, 30; au Pistou, 28
Sour Cream: Dressing, Celery Cabbage

Sour Cream (cont.)
with, 169; Sauce, Summer Broccoli Ring with, 89
South American dishes: Brazilian Black Beans, 124; Latin American Vegetable Stew, 64; Peruvian Mixed Vegetable Soup, 30; Peruvian Potato Salad with Cottage Cheese, 147; Peruvian Sweet-Potato Appetizers, 11; South American Corn Pudding, 138; South American Eggplant-Filled Green Peppers, 51; Venezuelan Pumpkin Soup, 22; Venezuelan Spinach-Avocado Salad, 151
Southern: Hoppin' John, 123; Sweet Potato and Nut Cake, 181
South Pacific Mixed Salad, 151
South Seas Stuffed Sweet Potatoes, 53
Southwestern Frijoles, 118
Soy and Whole Wheat Muffins, 176
Soyburgers, 125
Spanish dishes: Gazpacho from Andalusia, 26; Omelette Basque, 77; Spanish Stuffed Tomatoes, 50
Spinach: Japanese with Sesame, 192; -Mushroom Soup, Hong Kong, 27; Pie, Greek, 92; Ring, with Creamed Eggs and Walnuts, 83; Roman Style, 97; Salad, Venezuelan Avocado-, 151 -Zucchini Casserole, California, 88
Stew: Istanbul Vegetable, 72; Latin American Vegetable, 64; Virginia Cress, 108
Sukiyaki, Japanese Vegetable, 62
Summer Broccoli Ring with Sour Cream Sauce, 89
Swedish: Mashed Rutabaga and Potatoes, 137; Vegetable-Macaroni Salad, 153; Vegetable-Rice Pot, 65
Sweet Potato(es), 11; Appetizer, Peruvian, 11; -Carrot Tzimmes, Israeli, 140; -Coconut Pie, Louisiana, 184; Custard, Maryland, 178; and Nut Cake, Southern, 181; Pudding, Puerto Rican, 181; Rolls, Georgia, 176; South Sea Stuffed, 53
Sweet-Sour Vegetable Relish, 187
Swiss Chard, Puerto Rican, 102
Swiss Onion Tart, 80

Tart, Swiss Onion, 80

Taverna Beans, Turkish, 116
Tempura, Vegetable, from Japan, 17
Timballo, Italian Macaroni-Zucchini, 86
Tomato(es), 50; -Artichoke Cocktail, California, 11; -Bread Salad, 148; Chutney, Pakistan, 191; Flower Salad, 143; Fried Green, with Milk Gravy, 128; Greek Rice-Stuffed, 53; -Herb Whole Wheat Bread, 172; Marmalade, 194; Ketchup, Old-Fashioned, 199; Oregon Salmon-Stuffed, 54; Pie, Hoosier Green, 180; Relish, Green, 187; -Rice Gumbo, Creole, 35; Salad Ring with Cucumbers, 150; Sauce, Basic, 195; Spanish Stuffed, 50; -Zucchini Lasagna, 63
Tuna Fish: Italian White Beans with, 117; -Stuffed Green Peppers, Florida, 39
Turkish: Cold Green Beans, 132; Cauliflower Salad, 143; Lamb-Rice Stuffed Eggplant, 44; Miniature Potato Balls, 13; Pumpkin-Nut Pudding, 179; Taverna Beans, 116
Turnip(s): 19, 45; Custard, Old-Fashioned, 130; Greens and Pot Likker, Alabama, 106; Nebraska Stuffed, 45. *See also* Swedish Mashed Rutabaga and Potatoes.
Tuscan-Bean Pasta Soup, 32

Venetian Risi e Bisi, 131
Venezuelan: Pumpkin Soup, 22; Spinach-Avocado Salad, 151

Vermont: Acorn Squash, 134; Mess o' Greens with Cornmeal Dumplings, 105
Viennese Wilted Cucumbers, 146
Virginia Cress Stew, 108
Virgin Island Stuffed Chayotes, 168

Walnut-Pumpkin Bread, 174
Watercress Soup, Parisian, 107
Wax Beans, Creole, 139
White Beans, 32, 116; Italian, with Tuna Fish, 117; Picnic Herbed, 121; Yugoslavian Cold, 121
White Radish(es), Japanese, with Sesame, 166; Oriental, Medley, 167
Whole Wheat: Bread, Tomato-Herb, 172; -Soy Muffins, 176
Winter Vegetable Soup from Russia, 21

Yam-Coconut Pie, Louisiana, 184
Yogurt: Dip, Balkan Raw Vegetable, 16; Sauce, Lebanese Stuffed Squash with, 43; Soup, Circassian Vegetable-, 31; Soup, Iranian Cold Cucumber, 25
Yugoslavian Cold White Beans, 121

Zucchini, 7; Casserole, California Spinach-, 88; Lasagna, Tomato-, 63; Mediterranean Stuffed, 48; Mexican Corn- Stuffed, 40; Quiche au Fromage, 7; Relish, California, 189; Timballo, Italian Macaroni-, 86